SpringerBriefs in Physics

SpringerBriefs in Physics are a series of slim high-quality publications encompassing the entire spectrum of physics. Manuscripts for SpringerBriefs in Physics will be evaluated by Springer and by members of the Editorial Board. Proposals and other communication should be sent to your Publishing Editors at Springer.

Featuring compact volumes of 50 to 125 pages (approximately 20,000–45,000 words), Briefs are shorter than a conventional book but longer than a journal article. Thus, Briefs serve as timely, concise tools for students, researchers, and professionals.

Typical texts for publication might include:

- A snapshot review of the current state of a hot or emerging field
- A concise introduction to core concepts that students must understand in order to make independent contributions
- An extended research report giving more details and discussion than is possible in a conventional journal article
- A manual describing underlying principles and best practices for an experimental technique
- An essay exploring new ideas within physics, related philosophical issues, or broader topics such as science and society

Briefs allow authors to present their ideas and readers to absorb them with minimal time investment.

Briefs will be published as part of Springer's eBook collection, with millions of users worldwide. In addition, they will be available, just like other books, for individual print and electronic purchase.

Briefs are characterized by fast, global electronic dissemination, straightforward publishing agreements, easy-to-use manuscript preparation and formatting guidelines, and expedited production schedules. We aim for publication 8–12 weeks after acceptance.

More information about this series at http://www.springer.com/series/8902

Irinel Caprini

Functional Analysis and Optimization Methods in Hadron Physics

 Springer

Irinel Caprini
Department of Theoretical Physics
National Institute of Physics and Nuclear
Engineering
Bucharest-Magurele, Romania

ISSN 2191-5423   ISSN 2191-5431   (electronic)
SpringerBriefs in Physics
ISBN 978-3-030-18947-1   ISBN 978-3-030-18948-8   (eBook)
https://doi.org/10.1007/978-3-030-18948-8

This Springer imprint is published by the registered company Springer Nature Switzerland AG
The registered company address is: Gewerbestrasse 11, 6330 Cham, Switzerland

Preface

Consequences of the fundamental properties of causality and unitarity, the analytic properties of the scattering amplitudes and Green functions have a long and interesting history in particle physics. The lack of a consistent quantum field theory for the strong interactions in the 1960s led to the bold speculation that analyticity might be a viable substitute for a quantum field theory. This approach was abandoned after the development of the Standard Model (SM) of particle physics, in particular after the creation of Quantum Chromodynamics (QCD), the modern theory of strong interactions.

The interest in analyticity increased again in recent years, when it became clear that the strong interactions, which through radiative corrections affect all the processes, including the electroweak ones, are at present the biggest source of theoretical uncertainty in the tests of the SM and the search for new physics signals. While the typical errors of many pure electroweak observables are of the order of 1–2%, those related to the strong interactions often reach 10% or more, especially for observables where perturbative QCD cannot be applied. In this context, along with Chiral Perturbation Theory (ChPT) and lattice QCD, analyticity proves to be a useful tool for improving the precision of the theoretical predictions in hadron physics. Moreover, it turns out that more sophisticated mathematical methods, belonging to functional analysis and optimization theory, are often required in order to properly treat the information available for the physical quantities of interest.

These methods prove to be helpful for both ways of exploiting analyticity in physics, namely dispersion relations and expansions in power series. In practical applications, it may happen that the physical input available for a scattering amplitude or form factor does not allow the straightforward application of the standard dispersion relations, based on the Cauchy integral. In this case, more powerful techniques can be useful, avoiding the need for ad hoc model-dependent assumptions. With methods of optimization theory and functional analysis, one can also improve the convergence of power series, and even provide rigorous estimates of the truncation error. The present book aims to offer a systematic presentation of the mathematical background and of some physical problems where these techniques prove to be useful.

The book has six chapters. Chapter 1 contains a brief historical overview of the standard dispersion relations in particle physics. The status of analyticity within the Standard Model is discussed in Chap. 2 where the need for methods of functional analysis and optimization theory is illustrated on several examples. Chapter 3 contains a selection of mathematical results from complex and functional analysis, which are applied in Chap. 4 for solving several specific optimization problems encountered in hadron physics. This chapter, the most extended of the book, offers a unified presentation of problems and techniques that can be found otherwise only scattered in many places in the literature. A brief review of the most important physical applications is also given here. The last two chapters are focused on two specific applications: Chap. 5 presents in a historical perspective the derivation of model-independent constraints on several weak and electromagnetic hadronic form factors and Chap. 6 deals with the application of functional analysis and optimization methods in perturbative QCD. In each case, I emphasize the physical problem of interest and the mathematical technique used for solving it, sending for details to the original articles. I hope that the book will be of help to researchers and graduate students working or interested in precision predictions in hadron physics.

Acknowledgements

I would like to acknowledge the pioneering work on analytic continuation and optimization theory, performed in the early 1970s by the Bucharest particle theory group led by Sorin Ciulli, where I became familiar with many of the mathematical techniques presented in this book. I thank B. Ananthanarayan, Diogo Boito, Claude Bourrely, Jan Fischer, Maarten Golterman, Laurent Lellouch, Matthias Neubert and Santiago Peris for fruitful collaborations on the application of functional analysis and optimization methods to hadron physics.

Bucharest, Romania   Irinel Caprini
March 2019

Contents

Chapter 1
Theory of Strong Interactions Before the Standard Model

In this chapter we briefly review the attempts to construct a theory of the strong interactions in the 1960s, when it became obvious that a perturbatively renormalizable quantum field theory of these forces was not possible. The derivation of the analytic properties of scattering amplitudes and form factors from the general principles of causality and unitarity in quantum field theory is discussed at a qualitative level, using the Lehmann–Symanzik–Zimmermann (LSZ) formalism. Standard dispersion relations based on Cauchy integral formula are written down for scattering amplitudes and form factors. Finally, some pitfalls in practical applications, related to the so-called instability of analytic continuation, are emphasized.

1.1 Strong Interactions and Quantum Field Theory

After the impressive success of the perturbatively renormalized Quantum Electrodynamics (QED) at the beginning of 1950s, a similar theory was sought for the strong and weak nuclear interactions. However, the task proved to be highly nontrivial. For the strong interactions there were two types of difficulties. First, the "coupling constant" of these forces is large, so it was hard to imagine how one can approximate the full result with a few terms in a perturbation expansion. Secondly, it became clear that the proton, the neutron, the π meson, and all the particles able to interact by strong forces (denoted generically as "hadrons") are composite, extended objects, and not elementary particles as required by a field theory like QED. Hence, in the 1960s, the hopes to build a quantum field theory for the strong interactions diminished dramatically.

In this context, an alternative view based on a different ideology started to gain ground. The starting remark was that actually very few observables are accessible in high energy experiments. Typically, in a high-energy collision of two initial particles, only the final stable particles produced in the collision are measured. Therefore, it was reasonable to assume that the fundamental object is the scattering matrix (S-matrix), an operator introduced by Heisenberg, which connects the input and

© The Author(s), under exclusive licence to Springer Nature Switzerland AG 2019

I. Caprini, *Functional Analysis and Optimization Methods in Hadron Physics*, SpringerBriefs in Physics, https://doi.org/10.1007/978-3-030-18948-8_1

output of a scattering experiment without seeking to give a localized description of the intervening events. This approach retained from quantum field theory (QFT) only the general principles valid in any model: relativistic invariance, causality and unitarity, i.e. conservation of probability in the Hilbert space of states. By suitably exploiting them, it was possible to prove that the S matrix is an analytic[1] complex function of the relativistic invariants constructed from the momenta of the external particles, with singularities (poles or branch points) having a physical significance. This property allows one to analytically continue the scattering amplitudes outside the physically-accessible regions, into unphysical regions. Moreover, the property called "crossing symmetry" implied that the same analytic function describes not only a certain process, but also the related processes obtained by permutting one incoming and one outgoing particle. In this frame, the particles stable with respect to strong interactions are associated to poles situated on the real axis in the complex energy plane, while the unstable particles, the so-called resonances, are associates to complex poles on the higher Riemann sheets of this plane, and the only other singularities of the scattering amplitudes are branch cuts generated from the contribution of the stable states in the unitarity sum.

Many phenomenological applications of this approach, whose most radical formulation is known as the "analytic S-matrix theory" [1], have been performed. The aim was to correlate in a consistent picture the stable and unstable hadrons, found by analytic extrapolation from the physical regions to unphysical points in the complex energy plane. A remarkable property in complex analysis, namely the uniqueness of analytic extrapolation, was considered to be a strong point in favour of the picture. Thus, it is not surprising that the seminal book [1] started with the arch statement: "One of the most important discoveries in elementary particle physics has been that of the existence of the complex plane"!

Actually, not very much from complex analysis was exploited in physical investigations: most of the practical applications have been based on the Cauchy integral relation, which expresses an analytic function in a domain of the complex plane in terms of its values on the boundary of the domain. Combined with a symmetry known as "Schwarz reflection" shown to be satisfied by the scattering amplitudes, Cauchy relation led to the famous "dispersion relations", which relate the real part of an amplitude to its imaginary part. The usefulness of these relations was obvious especially for amplitudes of forward scattering, where the imaginary part is related by unitarity to the total cross section measured experimentally. Dispersion relations were used not only for the total amplitudes, but also for its projections on angular momentum, the so-called partial wave amplitudes, especially in order to detect resonances of definite spin and parity. Also, the consequences of analyticity have been extended to other quantities, like the form factors measured in the electromagnetic and weak interactions of hadrons. In the next two sections we will briefly review the analyticity properties of the scattering amplitudes and hadronic form factors.

---

[1] We shall use both terms "analytic" and "holomorphic", which are equivalent for complex-valued functions of complex variables.

1.2 Analytic Properties of Scattering Amplitudes

In this section we first introduce the S matrix and the scattering amplitude and write down the consequences of unitarity for total and partial wave amplitudes. We then present the hints on analyticity obtained from the study of Feynman integrals and the more rigorous methods based on axiomatic field theory, using the Lehmann–Symanzik–Zimmermann (LSZ) formalism. The most famous rigorous results on the asymptotic behavior at large energies are also given. We then discuss the analytic properties of the partial waves and the continuation into the second Riemann sheet, and finally mention for completeness the analyticity in the Lehmann ellipse.

We consider the scattering process

$$a(p_a) + b(p_b) \rightarrow c(p_c) + d(p_d), \tag{1.1}$$

represented in Fig. 1.1, where a, b, c, d are hadrons and $p = (p_0, \mathbf{p})$ with $p^2 = p_0^2 - \mathbf{p}^2$ are 4-momenta which satisfy momentum conservation

$$p_a + p_b = p_c + p_d. \tag{1.2}$$

For simplicity, we assume that the particles are spinless, of zero isospin and of equal masses. The Lorentz invariants constructed from the momenta define the Mandelstam variables

$$s = (p_a + p_b)^2, \quad t = (p_a - p_b)^2, \quad u = (p_a - p_c)^2. \tag{1.3}$$

Using the mass-shell relations $p_i^2 = m^2$ for $i = a, b, c, d$, one obtains the relation

$$s + t + u = 4m^2, \tag{1.4}$$

which implies that the process is described by two independent Lorentz invariants. In the centre-of-mass system (c.m.s) $\mathbf{p}_1 + \mathbf{p}_2 = \mathbf{0}$, the variables s and t are the total energy squared and the momentum transfer, respectively, expressed as

$$s = 4(q^2 + m^2), \quad t = -2q^2(1 - \cos\theta), \tag{1.5}$$

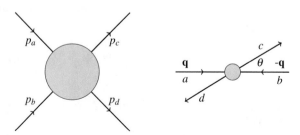

Fig. 1.1 Elastic scattering of two hadrons. Right: scattering in the center of mass system. The blob indicates the strong dynamics

where $q = |\mathbf{q}|$ is the magnitude of the initial momentum in c.m.s. and θ is the scattering angle between particles a and c in this system. The physical region of the process (1.1), denoted as s-channel, is $s \geq 4m^2$ and $t \leq 0$.

The S-matrix is defined as

$$S_{fi} = \langle c(p_c), d(p_d); \text{ out}|a(p_a), b(p_b); \text{ in}\rangle, \tag{1.6}$$

where "in" and "out" are complete set of states at $t \to -\infty$ and $t \to \infty$, respectively. It is written as

$$S_{fi} = \delta_{fi} + (2\pi)^4 i \delta^{(4)}(p_a + p_b - p_c - p_d) T_{fi}, \tag{1.7}$$

where the first term denotes the non-scattering part and T is the scattering amplitude, which depends only on two independent Mandelstam variables $T_{fi} \equiv T(s, t)$.

Unitarity of the S matrix, fundamental to the probabilistic interpretation of quantum theory,

$$S^\dagger S = SS^\dagger = 1 \tag{1.8}$$

implies for the scattering amplitude the relation

$$i(T^*_{if} - T_{fi}) = \sum_n (2\pi)^4 \delta^{(4)}(p_i - p_n) T^*_{nf} T_{ni}. \tag{1.9}$$

In particular, for the forward scattering $\theta = 0$, which implies $t = 0$, this relation gives the so-called optical theorem

$$\text{Im} T_{ii} = \frac{1}{2} \sum_n (2\pi)^4 \delta^{(4)}(p_i - p_n)|T_{ni}|^2, \tag{1.10}$$

where the sum in the r.h.s. is proportional to the total cross section of the process $a + b \to \text{anything}$.

By expanding the amplitude in a series of Legendre polynomials

$$T(s, t) = \sum_{l \geq 0} (2l + 1) t_l(s) P_l(\cos\theta), \tag{1.11}$$

one obtains the partial waves $t_l(s)$ of definite angular momentum in the s-channel by the projection

$$t_l(s) = \frac{1}{2} \int_{-1}^{1} dz\, T(s, -2q^2(1 - z))\, P_l(z). \tag{1.12}$$

One can further show that the unitarity condition (1.9) can be written in terms of $t_l(s)$ as

$$\text{Im}\, t_l(s) = \rho(s) t_l^*(s) t_l(s) \theta(s - 4m^2) + \cdots \tag{1.13}$$

where $\rho(s)$ is an invariant phase-space factor. At low energies, below the first "inelastic" threshold s_{in}, when only the contribution of states of two particles of mass m (assumed to be the lowest mass of hadrons in nature) is allowed in the sum (1.9), the r.h.s. of (1.13) reduces to the first term, and the solution of this equation is parametrized most generally as

$$t_l(s) = \frac{e^{2i\delta_l(s)} - 1}{2i\rho(s)}, \qquad s < s_{in}, \tag{1.14}$$

in terms of the real phase shift $\delta_l(s)$. At higher energies, the most general expression of t_l is

$$t_l(s) = \frac{\eta_l(s)e^{2i\delta_l(s)} - 1}{2i\rho(s)}, \tag{1.15}$$

where $\eta_l(s) \leq 1$, denoted as "elasticity", accounts for the contribution of inelastic states. We note also that the S-matrix projection $S_l(s)$, defined in terms of the amplitude by

$$S_l(s) = 1 + 2i\rho(s)t_l(s), \tag{1.16}$$

is parametrized for $s \geq 4m^2$ as

$$S_l(s) = \eta_l(s)e^{2i\delta_l(s)}, \tag{1.17}$$

and satisfies the condition $|S_l(s)| \leq 1$.

Another important concept in particle physics in the 1960s was known as "crossing symmetry". It means that the amplitude $T(s, t)$ can be analytically continued from the physical region $s \geq 4m^2$ and $t \leq 0$ of the process (1.1), denoted as "s-channel", to complex values of the variables. Moreover, the same analytic function describes also the "crossed channels"

$$\begin{aligned} a(p_a) + \bar{c}(-p_c) &\to \bar{b}(-p_b) + d(p_d), \\ a(p_a) + \bar{d}(-p_d) &\to c(p_c) + \bar{b}(-p_b), \end{aligned} \tag{1.18}$$

when the variables reach physical values where these processes, denoted as t-channel and u-channel, respectively, can occur (here $\bar{b}$ denotes the antiparticle of b, etc.). This could actually be used in dynamical calculations, relating bound states in one channel to forces in the other. Analytic continuation upon the angular momentum l of the partial wave amplitudes has been also considered, leading to useful results about the asymptotic behavior of the scattering amplitudes at large energies, within the so-called Regge theory. These concepts reshaped much of theoretical research on strong interaction physics throughout the 1950 and 1960s.

There are several types of arguments leading to the conclusion that the scattering amplitudes are analytic functions of the complex Lorentz invariants constructed from the particle momenta. Nonrelativistic quantum mechanics offered useful insights especially for the analytic continuation of the angular momentum variable. A solid ground was provided also by perturbative expansions in QFT. Although perturbation theory was not believed to provide a quantitative description of hadronic processes, the inspection of special low-order Feynman diagrams showed that the amplitudes had poles and branch points in the complex planes of the kinematical variables. While a general proof of analyticity based on perturbative QFT is lacking, these investigations led to a deeper understanding of the singularities of Feynman integrals [2, 3].

These singularities can be found by using a lemma of Hadamard, which considers the singularities of a function defined by an integral representation like

$$g(z) = \int_C F(w, z)dw, \qquad (1.19)$$

where $F(w, z)$ is a function of two complex variables and C is a contour in the complex w plane. Clearly, if the integral converges for a value z_0 of z, the function $g(z)$ will be holomorphic at z_0. Problems can occur when a singularity $w = \zeta(z)$ of $F(w, z)$ reaches for a certain z_1 the integration contour C. Then, on may think that the function $g(z)$ will not be holomorphic at $z = z_1$. However, if it is possible to deform the integration contour such that to avoid the singularity, the function will be still holomorphic at $z = z_1$. This is illustrated with the point P_1 in Fig. 1.2.

There are nevertheless two situations when this deformation is not possible: one occurs when two singularities $w = \zeta_2(z)$ and $w = \zeta_3(z)$ of $F(w, z)$, moving in the complex w plane when z varies, meet for a certain $z = z_p$ at the same point P of the contour C coming from opposite sides of the contour (see Fig. 1.2). Then the corresponding point of the contour is "pinched" and cannot be displaced. As a consequence, the function $g(z)$ will have a singularity (denoted as "pinch singularity") at $z = z_p$. The other case, shown in Fig. 1.2, occurs when the contour C is open and a singularity $w = \zeta_4(z)$ of $F(w, z)$ reaches for a certain $z = z_e$ one end point of C. Then the contour can no longer be deformed and $z = z_e$ will be a singularity for $g(z)$. The singularities generated by this situation are called "end-point singularities". We recall that in the context of Feynman integrals, the singularities generated in the way described above are known as Landau singularities. For any Feynman integral written in parametric form, there exists a discriminant $D(p_i \cdot p_j, m_k^2, \alpha_k)$ depending on the Lorentz invariants constructed from the momenta and a set of parameters α_k, such that the singularities of the integral are given by

Fig. 1.2 Types of singularities in the w-complex plane. The contour between the points A and B can be deformed so as to avoid the simple pole P_1, but this is not true for the pinch (P_2, P_3) and end-point (P_4) singularities

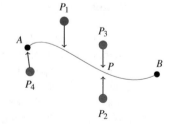

$$\partial D/\partial \alpha_k = 0, \quad k = 1, 2 \ldots \tag{1.20}$$

The investigation of the poles and branch points of the Feynman integrals offered insights on the singularities of the scattering amplitudes, not a rigorous proof of the analytic properties. The adequate framework for the rigorous derivation of analyticity of the S-matrix proved to be the axiomatic formulation of QFT [4–6]. This approach exploits the basic principles of QFT: relativistic invariance, causality (the fact that no signal can propagate faster than light) and unitarity (conservation of probability), supplemented by a few very general assumptions inspired from the physics of the scattering processes. These principles are implemented in a way convenient for the proof of analyticity in the Lehmann–Symanzik–Zimmermann (LSZ) reduction formalism [5, 6] (see also [7] for a clear presentation without technical details).

The axiomatic approach defines a set of fields, denoted as $\phi_{\text{in}}(x)$ and $\phi_{\text{out}}(x)$, associated to the free states "in" and "out", and assumes the existence of an "interpolation field" $\phi(x)$, which converges (in a weak sense, i.e. in matrix elements) to the "in" and "out" fields for $t \to -\infty$ and $t \to \infty$, respectively. So, one can write

$$\langle |\phi_{\substack{\text{in} \\ \text{out}}}(x)| \rangle = \lim_{t \to \mp\infty} Z^{-1/2} \langle |\phi(x)| \rangle \tag{1.21}$$

where Z is a renormalization constant, whose value is not relevant here.

The LSZ formalism expresses the S-matrix element (1.6) in terms of the interpolation fields of the initial and final particles in the scattering process. In order to do this, we first "reduce" the particle $a(p_a)$ in the state $|a(p_a), b(p_b); \text{in}\rangle$, by writing

$$|a(p_a), b(p_b); \text{in}\rangle = a^\dagger_{\text{in},a} |b(p_b)\rangle, \tag{1.22}$$

where $a^\dagger_{\text{in},a}$ is the creation operator of the "in" particle a, written as

$$a^\dagger_{\text{in},a} = -i \int \frac{d^3 x}{\sqrt{(2\pi)^3 \, 2\omega_{p_a}}} e^{-ip_a x} \overset{\leftrightarrow}{\partial_0} \phi_{\text{in},a}(x), \tag{1.23}$$

in terms of the operator $\phi_{\text{in},a}$, where $\omega_p = \sqrt{\mathbf{p}^2 + m^2}$ is the energy of the particle.

By applying the above reduction, after some standard manipulations [7] and leaving aside terms that can contribute only to the "nonscattering" part δ_{fi} of (1.7), we write the matrix element (1.6) as

$$Z^{-1/2} \int dx \frac{e^{-ip_a \cdot x}}{\sqrt{(2\pi)^3 \, 2\omega_{p_a}}} \mathcal{K}_x \langle c(p_c), d(p_d); \text{out}|\phi_a(x)|b(p_b)\rangle, \tag{1.24}$$

where $dx = dx_0 \, d^3x$ and $\mathcal{K}_x = \partial_x^2 + m^2$ is the Klein–Gordon operator.

After the similar LSZ reduction of the final particle c, we obtain

$$S_{fi} = \delta_{fi} + \frac{Z^{-1}}{\sqrt{2\omega_c\, 2\omega_a}}\frac{1}{(2\pi)^3} \tag{1.25}$$

$$\times \int dx \int dy \, e^{ip_c \cdot y - ip_a \cdot x} \mathscr{K}_x \mathscr{K}_y \langle d(p_d)|\theta(y_0 - x_0)[\phi_a(x), \phi_c(y)]|b(p_b)\rangle.$$

The appearance of the commutator in the representation (1.25) is essential for analyticity, since it allows to implement causality. More exactly, the so-called "microcausality" property takes the form

$$[\phi(x), \phi(y)] = 0, \qquad (x - y)^2 < 0, \tag{1.26}$$

expressing the fact that for spacelike separations the fields are independent. Together with the function $\theta(x_0 - y_0)$ appearing in (1.25), the relation (1.26) allows the analytic continuation of the amplitude in the complex plane of the external 4-momenta p_a and p_c.

This result can be understood using a simple example: consider a function $\tilde{f}(p)$ defined by the integral

$$\tilde{f}(p) = \int_{-\infty}^{\infty} dx \, e^{ipx} f(x), \tag{1.27}$$

where $f(x)$ vanishes for $x < 0$, i.e. it can be written as $f(x) = \theta(x) f_1(x)$. Then the integral is performed only on $x > 0$ and it converges if p is complex with a positive imaginary part: indeed $ipx = ix\,\mathrm{Re}\,p - x\,\mathrm{Im}\,p$ has a negative real part for $x > 0$ and $\mathrm{Im}\,p > 0$. Therefore, the integral (1.27) defines a function $\tilde{f}(p)$ analytic in the upper half of the complex p plane.

In the case of the S matrix element (1.25), the principle is the same but the proof is technically more complicated due to the presence of many variables. The rigorous proofs of analyticity of the scattering amplitudes $T(s, t)$ in the axiomatic formalism [8–15] use as mathematical tool the so-called "edge-of-the-wedge theorem", which implies that holomorphic functions on two "wedges" with an "edge" in common are analytic continuations of each other, provided they are equal to the same continuous function on the edge. The difficulty of the proof is to identify two kinematical regions in terms of relativistic invariants (Mandelstam variables) where (1.25) holds, and to connect the two regions through an interval situated on the real axis, where a spectral representation given by unitarity holds.

In order to see in a heuristic how the spectral representation looks, we note that the expression the Heaviside step function

$$\theta(y) = \frac{1}{2\pi i} \int_{-\infty}^{\infty} dt \, \frac{e^{ity}}{t - i\varepsilon}, \qquad \varepsilon > 0, \tag{1.28}$$

implies that the S-matrix (1.25) has branch cuts with discontinuities obtained by formally replacing $i\theta(y_0 - x_0)$ by 1/2. This results from the formal Plemelj relation[2]

$$\frac{1}{z \pm i\varepsilon} = P\left(\frac{1}{z}\right) \mp i\pi\delta(z), \tag{1.29}$$

where P denotes the principal value.

By inserting a complete set of "in" or "out" states $|n\rangle$ in the commutator, the discontinuities of the S-matrix across the cuts in the complex s-plane at fixed t are proportional to

$$\int dx \int dy \, e^{iP_d \cdot y - iP_a \cdot x} \left[\sum_n \langle c|\eta_a(x)|n\rangle \langle n|\eta_d(y)|b\rangle - \sum_n \langle c|\eta_d(y)|n\rangle \langle n|\eta_a(x)|b\rangle \right],$$

where $\sum_n$ denotes the summation and integration over the momenta of all physical particles allowed by momentum conservation and

$$\eta(x) = \mathscr{K}_x \phi(x) \tag{1.30}$$

is the "source" of the interpolation field. By using the space-time translation invariance

$$\eta(x) = e^{iP \cdot x} \eta(0) e^{-iP \cdot x}, \tag{1.31}$$

performing the integrals and taking out the delta function which expresses momentum conservation, leads to an expression for the discontinuity of the amplitude $T(s, t)$ across the cuts in the s-plane at fixed t

$$\text{disc}|_s T(s, t) \equiv T(s + i\varepsilon, t) - T(s - i\varepsilon, t). \tag{1.32}$$

Actually, it can be shown that the amplitude satisfies an additional symmetry property named "Schwarz reflection", which stipulates that, at a fixed real t less than a certain t_0

$$T(s^*, t) = T^*(s, t), \tag{1.33}$$

where * denotes complex conjugation. From this property, it follows that the discontinuity (1.32) can be written in terms of the imaginary part of the amplitude above the cut:

$$\text{disc}|_s T(s, t) = 2i \text{Im} T(s + i\varepsilon, t). \tag{1.34}$$

Therefore, unitarity implies that the imaginary part of the amplitude is expressed as

---

[2]This relation is known also as Sokhotski–Plemelj, or Plemelj–Privalov relation [16].

$$\text{Im}T(s + i\varepsilon, t) \sim \sum_n \delta(p_n - p_a - p_b)\langle d|\eta_c(0)|n\rangle\langle n|\eta_a(0)|b\rangle$$

$$- \sum_n \delta(p_n - p_b + p_c)\langle d|\eta_a(0)|n\rangle\langle n|\eta_c(0)|b\rangle, \qquad (1.35)$$

In the first sum contribute states $|n\rangle$ with $p_n = p_a + p_b$, which implies $p_n^2 = s$. The lowest contribution can be a one-particle state of mass M^2, which leads to an imaginary part of the form

$$\text{Im}T(s + i\varepsilon, t) \sim \delta(s - M^2), \qquad (1.36)$$

which shows that the amplitude itself has a pole

$$T(s, t) \sim \frac{g^2}{s - M^2}, \qquad (1.37)$$

where the residue is determined by the coupling g of the external particles to the state of mass M. The multiparticle states in the first sum in (1.35) produce a cut in the s-plane along the real axis for $s \geq s_R$, where s_R is the lowest invariant mass of the particles that can contribute. Traditionally, this cut is called "right-hand cut".

In the second sum contribute states with $p_n = p_b - p_c = p_d - p_a$, which implies $p_n^2 = u$. As before, a one-particle state of mass M in the u-channel is responsible for a pole in the s-plane, which, using (1.4), can be written as

$$T(s, t) \sim \frac{g^2}{s - 4m^2 + t + M^2}. \qquad (1.38)$$

The multiparticle states in the u-channel generate a cut in the s-plane, along the real axis for $s \leq 4m^2 - u - u_0$. This is called "left-hand cut".

From the above heuristic discussion, it follows that the generic analytic structure of $T(s, t)$ in the s-plane, for fixed t in a suitable interval, looks as shown in Fig. 1.3, with two cuts along the real axis, a right-hand cut for $s \geq s_R$ produced by unitarity in the direct, s-channel, and a left-hand cut for $s \leq s_L$ produced by the intermediate states in the crossed u-channel.

As mentioned above, it is not trivial to prove in a rigorous way these properties. In fact, as shown in [8–15], rigorous proofs of analyticity exist only for a few scattering amplitudes in the s-plane for $t \leq 0$. By an ingenious use of unitarity, the axiomatic

Fig. 1.3 Cuts of the amplitude $T(s, t)$ in the complex s-plane at fixed t

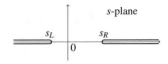

analyticity domain has been extended in [17] to a region in the complex t-plane including positive unphysical t.

In several cases it has been possible to obtain also rigorous bounds for the magnitude of the scattering amplitude at fixed t and large energies. For instance, the Froissart–Martin bound [18, 19] on the total cross section reads

$$\sigma_{\text{tot}}(s) \underset{s \to \infty}{<} C \, (\ln s/s_0)^2, \tag{1.39}$$

where C and s_0 are constant. In the particular case of $\pi\pi$ scattering, the constant C is calculable, $C = \pi/m_\pi^2$. Martin proved also a more general result

$$|T(s, t)|_{t \leq 0} \underset{s \to \infty}{<} C's \, (\ln s/s_0)^2, \tag{1.40}$$

where C' is a constant, and extended the validity of analyticity at fixed t to values of t greater than zero, and even complex, in a certain range of the t-plane. Upper and lower bounds on the differential cross sections have been obtained also by exploiting analyticity and unitarity [20].

The analytic properties of the total amplitude $T(s, t)$ have implications on the partial waves $t_l(s)$ defined by the projection (1.12). Using (1.5), this relation can be converted into an integral upon t. The partial wave amplitudes may have poles corresponding to stable (bound) states, and also right and left-hand cuts.[3] In particular, the partial-wave projection of a pole in the crossed channel, given by (1.38), produces a left-hand cut for $t_l(s)$ in the s-plane. Moreover, the partial waves $t_l(s)$ can be analytically continued beyond the first sheet of the complex s plane, into the second or higher Riemann sheets. This continuation is based on the unitarity relation (1.13), valid on the elastic part $4m^2 < s < s_{in}$ of the right-hand cut. Using Schwarz reflection property, which implies in particular that $t_l^*(s + i\varepsilon) = t_l(s - i\varepsilon)$, we write (1.13) as

$$\frac{t_l(s + i\varepsilon) - t_l(s - i\varepsilon)}{2i} = \rho(s)t_l(s + i\varepsilon)t_l(s - i\varepsilon), \qquad s > 4m^2. \tag{1.41}$$

We recall now that the first Riemann sheet is defined by $\arg(s - 4m^2) \in (0, 2\pi)$, while the second sheet is defined by $\arg(s - 4m^2) \in (2\pi, 4\pi)$. The amplitude $t_l^{\text{II}}(t)$ on the second sheet is defined by gluing the lower edge of the cut in the first sheet to the upper edge on the cut in the second sheet, i.e., by requiring $t_l^{\text{II}}(t + i\varepsilon) = t_l(t - i\varepsilon)$. Understanding all quantities without a superscript as defined on the first Riemann sheet, we write Eq. (1.41) as

$$t_l^{\text{II}}(s) = \frac{t_l(s)}{1 + 2i\rho(s)t_l(s)}. \tag{1.42}$$

---

[3] In the case of unequal masses, the partial waves have also a cut of circular shape in the complex s-plane.

The S matrix element, defined on the first sheet by (1.16), is written on the second sheet as

$$S_l^{II}(s) = 1 - 2i\rho(s)t_l^{II}(s).$$ (1.43)

Using the definition (1.42) of $t_l^{II}(s)$, one obtains:

$$S_l^{II}(s) = \frac{1}{S_l(s)}.$$ (1.44)

From this relation it follows that the poles of $t_l^{II}(s)$ [and of $S_l^{II}(s)$] correspond to zeros of $S_l(s)$ on the first sheet. A crucial result in strong dynamics is that complex poles on the second or higher Riemann sheets correspond to "resonances", i.e. particles unstable with respect to the strong interactions. From (1.44) it follows that an elastic resonance corresponds to a zero of the S matrix on the first Riemann sheet. We mention that rigorous results based on axiomatic field theory have been derived also for partial waves, one of the most famous being the lower bound $a^{00} > -1.75$ on the scattering length (the value of the partial wave t_0 at threshold) of the process $\pi^0\pi^0 \to \pi^0\pi^0$, derived in [21].

Another important result is that the total amplitude expressed by the partial wave expansion (1.11) can by analytically continued in the $z = \cos\theta_s$ complex plane, from the physical region in the s-channel, where $-1 \leq \cos\theta_s \leq 1$, to the physical regions of the crossed channels, where $|\cos\theta_s| > 1$. As shown by Lehmann [15], $T(s, \cos\theta_s)$ at fixed s is analytic inside an ellipse in the complex $\cos\theta_s$ plane with foci at ± 1 and an s-dependent semimajor axis determined by the the lowest branch-points allowed by unitarity for the process. Moreover, it has been shown that the imaginary part $\text{Im}\, T(s, \cos\theta_s)$ at fixed s is analytic in a larger Lehmann ellipse in the complex $\cos\theta_s$ plane [15, 17].

1.3 Analytic Properties of Form Factors

The hadronic form factors are quantities that appear in the description of the electromagnetic and weak interactions of hadrons. Treating these interactions to first order, one is led to study matrix elements of the electromagnetic or weak current operators $J_\mu(0)$ between vacuum and a hadronic state. These matrix elements, parametrized in terms of Lorentz-invariant form factors, are measured in electromagnetic and weak processes involving hadrons, and play an important role in the strong interaction dynamics.

In Sect. 1.3.1, using for illustration the pion electromagnetic form factor, we derive the analyticity properties of form factors in the frame of LSZ formalism, prove Fermi–Watson theorem and discuss the analytic continuation into the second Riemann sheet. We also present Goldberger–Treiman version of unitarity sum which preserves Schwarz reflection and make some comments on unphysical cuts and anomalous thresholds. In Sect. 1.3.2 we consider the $\omega\pi$ electromagnetic form factor and show why Schwarz reflection is violated in this case.

1.3.1 Electromagnetic Form Factor of the Pion

The pion electromagnetic form factor $F_\pi^V(t)$ describes the coupling of an off-shell photon to a pair of charged pions. It is defined by the matrix element

$$\langle \pi^+(p')\pi^-(p); \text{out}|J_\mu^{\text{elm}}(0)|0\rangle = (p'-p)_\mu F_\pi^V(t), \qquad (1.45)$$

where $J_\mu^{\text{elm}}(x)$ is the electromagnetic current operator and $t = q^2 = (p+p')^2$. This quantity has been investigated since the 1950s and is still of much interest at present for the precision tests of the Standard Model. It is known also as the pion vector form factor, to distinguish it from the scalar form factors defined as matrix elements of scalar operators.

We shall discuss the analyticity of the pion electromagnetic form factor following the treatment of the dispersion techniques in field theory given in [7]. The aim is to emphasize the specific features and to understand the differences which appear in the case of other form factors, discussed in Sect. 1.3.2. For simplicity, we shall use in this subsection the notation $F(t) \equiv F_\pi^V(t)$.

It is convenient to define, along with $F(t)$, also other two form factors. One of them, $K(t)$, is defined as in (1.45), with the difference that the "out" pion pair is replaced by an "in" one:

$$\langle \pi^+(p')\pi^-(p); \text{in}|J_\mu^{\text{elm}}(0)|0\rangle = (p'-p)_\mu K(t). \qquad (1.46)$$

We consider also the form factor $G(t)$ defined by

$$\langle \pi^+(p')|J_\mu^{\text{elm}}(0)|\pi^+(p)\rangle = (p'+p)_\mu G(t), \qquad (1.47)$$

where $t = (p'-p)^2$. In the LSZ formalism, the uniparticle "in" and "out" asymptotic states coincide, so it is not necessary to specify them in (1.47).

We shall use now the invariance of the one-particle states under simultaneous space and time reversal (the PT operator) and the fact that time reveral leads to complex conjugates of the matrix elements. Then we obtain from (1.47):

$$\langle \pi^+(p')|PT(PT)^{-1}J_\mu^{\text{elm}}PT(PT)^{-1}|\pi^+(p)\rangle = \langle \pi^+(p')|J_\mu^{\text{elm}}|\pi^+(p)\rangle^*, \quad (1.48)$$

which implies that

$$G(t) = G^*(t). \qquad (1.49)$$

We apply now the PT transformation to the matrix element (1.45), using the fact that under time reversal the "out" states become "in" states and the T operator implies in addition the complex conjugation of the matrix element. Then (1.45) becomes

$$\langle \pi^+(p')\pi^-(p); \text{out}|PT(PT)^{-1}J_\mu^{\text{elm}}PT(PT)^{-1}|0\rangle = \langle \pi^+(p')\pi^-(p); \text{in}|J_\mu^{\text{elm}}|0\rangle^*,$$

which implies

$$F(t) = K^*(t). \tag{1.50}$$

The values of t for which the relations (1.49) and (1.50) are valid will be clear below, after studying the above matrix elements within the LSZ reduction formalism.

Let us perform first the LSZ reduction to the final $\pi^+(p')$ state. Then the matrix element (1.45) becomes

$$(p' - p)_\mu F(t) = Z^{-1/2} \int dx \frac{e^{ip' \cdot x}}{\sqrt{(2\pi)^3 2\omega_{p'}}} \langle \pi^-(p)|\theta(x_0)[\eta_{\pi^+}(x), J_\mu^{\mathrm{elm}}(0)]|0\rangle,$$

where $\eta(x)$ is the source of the interpolation field defined in (1.30). Using arguments similar to those presented for amplitudes in Sect. 1.2, one can prove from this relation that $F(t)$ can be defined as an analytic function in the upper half plane $\mathrm{Im}\, t > 0$ (for details see [7]).

Similarly, by LSZ reducing the $\pi^+(p')$ state in the matrix element (1.46), we obtain

$$(p' - p)_\mu K(t) = Z^{-1/2} \int dx \frac{e^{ip' \cdot x}}{\sqrt{(2\pi)^3 2\omega_{p'}}} \langle \pi^-(p)|\theta(-x_0)[\eta_{\pi^+}(x), J_\mu^{\mathrm{elm}}(0)]|0\rangle.$$

The presence of the function $\theta(-x_0)$ in this relation allows one to prove that $K(t)$ can be analytically continued in the lower half-plane $\mathrm{Im}\, t < 0$.

We subtract now the last two relations and take into account (1.50), to obtain

$$(p' - p)_\mu \mathrm{Im}\, F(t) = \frac{Z^{-1/2}}{2i} \int dx \frac{e^{ip' \cdot x}}{\sqrt{(2\pi)^3 2\omega_{p'}}} \langle \pi^-(p)|[\eta_{\pi^+}(x), J_\mu^{\mathrm{elm}}(0)]|0\rangle,$$

$$\tag{1.51}$$

where we used the relations $F(t) - F^*(t) = 2i\,\mathrm{Im}\, F(t)$ and $\theta(x_0) + \theta(-x_0) = 1$. The above derivation illustrates the rule mentioned above of obtaining the discontinuity from the θ function.

We insert now a complete set of states $|n\rangle$ in the commutator. For the first term in the commutator, the state of lowest mass which can contribute is a pair of two pions. Taking the intermediate states to be "in" states, we write this term as

$$I_1 = \frac{Z^{-1/2}}{2i} \int dx \frac{e^{ip' \cdot x}}{\sqrt{(2\pi)^3 2\omega_{p'}}} \int \frac{d^3k_1}{\sqrt{(2\pi)^3 2\omega_{k_1}}} \int \frac{d^3k_2}{\sqrt{(2\pi)^3 2\omega_{k_2}}} \tag{1.52}$$

$$\langle \pi^-(p)|\eta_{\pi^+}(x)|\pi^+(k_1)\pi^-(k_2); \mathrm{in}\rangle \langle \pi^+(k_1)\pi^-(k_2); \mathrm{in}|J_\mu^{\mathrm{elm}}(0)|0\rangle,$$

We use now translation invariance (1.31), which allows the exact integration upon the 4-variable x, yielding the delta function $\delta(p' + p - k_1 - k_2)$. We note also that the first matrix element in (1.52) is related to the total amplitude of $\pi\pi$ elastic scattering.

Indeed, from the relation (1.24), after using (1.30) and (1.31) and performing the trivial integration upon dx, we can write

$$T(s,t) = \frac{Z^{-1/2}}{\sqrt{(2\pi)^3 \, 2\omega_{p'}}} \langle \pi^-(p) | \eta_{\pi^+}(0) | \pi^+(k_1) \pi^-(k_2); \text{in} \rangle, \tag{1.53}$$

where $T(s,t)$ is the amplitude of the scattering $\pi^+(k_1) + \pi^-(k_2) \rightarrow \pi^+(p') + \pi^-(p)$, for which we use the partial wave expansion (1.11). A slight complication arises here because the pion has isospin equal to unity. Using the fact that isospin is conserved in strong interactions, we shall define amplitudes of definite isospin of the two pion states in the s-channel, $T^I(s,t)$, with $I = 0, 1, 2$. Accordingly, the partial waves defined in (1.12) will be written as $t_l^I(s)$.

We also note the appearance in (1.52) of the matrix element (1.46), which we express in terms of the form factor $K(t) = F^*(t)$. Then the integration upon the remaining variables is standard, leading to the relation

$$\text{Im}\, F(t) = \rho(t)\, t_1^1(t)\, F^*(t)\, \theta(t - 4m_\pi^2), \tag{1.54}$$

where $\rho(t) = \sqrt{1 - 4m_\pi^2/t}$ is the phase space.

Before analyzing further this relation, let us look at the second term of the commutator in (1.51):

$$\langle \pi^-(p) | J_\mu^{\text{elm}}(0) | n \rangle \, \langle n | \eta_{\pi^+}(x) | 0 \rangle. \tag{1.55}$$

In this case, the lowest state $|n\rangle$ that can contribute in the first matrix element is a one-pion state, giving (1.47). But the second matrix element vanishes

$$\langle \pi^- | \eta_{\pi^+}(x) | 0 \rangle = \mathcal{K}_x \langle \pi^- | \phi_{\pi^+}(x) | 0 \rangle = 0 \tag{1.56}$$

since the field is linear in creation and annihilation operators. Therefore, the imaginary part of the form factor $F(t)$ above the first threshold $4m_\pi^2$ imposed by unitarity is given by the states contributing only to the first unitarity sum in (1.51). The branch points produced by the physical states in the unitarity sum are known as normal or unitarity branch points.

We now come back to (1.54) and use the expression (1.15) of the partial wave, noting that in the elastic region $\eta_l^I(t) = 1$. Then (1.54) becomes

$$\frac{F(t) - F^*(t)}{2i} = \frac{e^{2i\delta_1^1(t)} - 1}{2i} F^*(t), \tag{1.57}$$

which implies that for t real in the elastic region, $4m_\pi^2 \leq t < t_{in}$, the following relation holds:

$$F(t) = F^*(t)\, e^{2i\delta_1^1(t)}. \tag{1.58}$$

Fig. 1.4 Analytic structure
of the pion vector form factor
in the complex t-plane

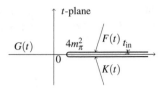

Writing the complex function $F(t)$ for real $t \geq 4m_\pi^2$ in terms of its modulus and phase, $F(t) = |F(t)|e^{i\varphi(t)}$, the relation (1.58) implies the equality

$$\varphi(t) = \delta_1^1(t) \text{ modulo } k\pi, \quad 4m_\pi^2 \leq t < t_{in}. \tag{1.59}$$

It is useful to emphasize that, since $F(t)$ is defined as an analytic function in the upper half of the t plane, by t in this relation we mean actually $t + i\varepsilon$ with $\varepsilon > 0$.

The relation (1.59) is the famous Fermi–Watson theorem [22, 23], stating that the phase of the pion electromagnetic form factor is equal (up to a multiple of π) to the phase shift of the elastic $\pi\pi$ partial-wave amplitude with $I = 1$ and $l = 1$. In (1.59) we emphasized the fact that the equality is valid at the energies where the only intermediate states acceptable by energy-momentum conservation in the unitarity sum consist from two pions. This region was named "elastic region" below (1.14) and extends up to the threshold t_{in} of creation of states consisting from four pions. With a good approximation this state can be identified to the $\omega\pi$ state.

From the above relations it follows that the three functions $F(t)$, $K(t)$ and $G(t)$ can be viewed as branches of a single analytic function in the complex t-plane (see Fig. 1.4). This function, which we shall denote as $F(t)$, is analytic in the complex t-plane cut along the real axis for $t > 4m_\pi^2$ and satisfies the Schwarz reflection property[4]

$$F(t^*) = F^*(t). \tag{1.60}$$

This implies that $F(t)$ is real for real $t < 4m_\pi^2$. In particular, for $t = 0$ charge conservation gives

$$F(0) = 1. \tag{1.61}$$

The relation (1.54), which is valid for t on the cut in the elastic region $4m_\pi^2 \leq t < t_{in}$, provides the route for the analytic continuation of the form factor onto the second Riemann sheet. We define, by analogy with the partial wave definition (1.42),

$$F^{II}(t + i\varepsilon) \equiv F(t - i\varepsilon), \tag{1.62}$$

and obtain from (1.54):

---

[4]Functions satisfying this property are sometimes said to be "analytic functions of real type", or "real-analytic functions", not to be confused with the real functions defined only on the real axis.

$$F^{\mathrm{II}}(t) = \frac{F(t)}{1 + 2i\rho(t)t_1^1(t)} = \frac{F(t)}{S_1^1(t)}. \tag{1.63}$$

Assuming that $F(t)$ does not vanish at the zeros of $S_1^1(t)$, $F^{\mathrm{II}}(t)$ has poles at that positions. So, the second-sheet poles of the form factor and the S-matrix element have the same positions, a known universality property of the poles in S-matrix theory. The relation (1.63) shows also that the analytic structure of the function $F^{\mathrm{II}}(t)$ is more complicated that that of $F(t)$: besides the unitarity cut, it has all the branch points of $S_1^1(t)$, in particular those lying on the left-hand cut produced by crossed-channel exchanges [7].

For t larger than the inelastic threshold t_{in}, when there are additional contributions to the unitarity sum (1.52), the relation (1.54) becomes

$$\mathrm{Im}\, F(t) = \rho(t)\, t_1^1(t)\, F^*(t)\, \theta(t - 4m_\pi^2) + \sigma_{in}(t)\, \theta(t - t_{in}), \tag{1.64}$$

with an inelastic term $\sigma_{in}(t)$ produced by states $|n\rangle$ of higher invariant mass in the sum (1.52). At first sight, a problem may appear: while the l.h.s. of this relation must be real, the r.h.s. may be no longer real. Indeed, in the inelastic region, the elasticity $\eta_1^1(t)$ in the expression (1.15) is less than 1, and the phases of $t_1^1(t)$ and $F(t)$ do not compensate each other in the product $t_1^1 F^*$. This should be true also for the product $\langle \pi^- | \eta_{\pi+} | n \rangle \langle n | J_\mu^{\mathrm{elm}} | 0 \rangle$ in the case of higher-mass states $|n\rangle$.

A solution to this apparent contradiction was found by Goldberger and Treiman [24], using the equivalence of the complete sets of "in" and "out" states. The first term in the r.h.s. of (1.64) appeared by inserting in the unitarity sum a complete set of $|\pi^+\pi^-;$ in$\rangle$ states. It is easy to see that if one would use the set $|\pi^+\pi^-;$ out$\rangle$, the first term in the r.h.s. of (1.64) would be replaced by $\rho(t)\, (t_1^1(t))^*\, F(t)$.

From this observation, the idea of Goldberger and Treiman was to write the sum over the complete set as

$$\sum_n |n\rangle\langle n| \;\rightarrow\; \frac{1}{2}\sum_n |n; \text{in}\rangle\langle n; \text{in}| + \frac{1}{2}\sum_n |n; \text{out}\rangle\langle n; \text{out}|. \tag{1.65}$$

It is easy to see that this trick transforms (1.64) into

$$\mathrm{Im}\, F(t) = \rho(t)\, \mathrm{Re}\, [t_1^1(t)\, F^*(t)]\theta(t - 4m_\pi^2) + \mathrm{Re}\, [\sigma_{in}(t)]\theta(t - t_{in}), \tag{1.66}$$

consistent with the reality of the l.h.s. for all values of t. Below the inelastic threshold t_{in}, the first term is already real by Fermi–Watson theorem and the application of the procedure is not necessary.

We end this section with two remarks, which are not actually relevant for the pion vector form factor, but refer to situations encountered in the case of other form factors and scattering amplitudes. The first situation appears when the invariant mass of the lowest state entering the unitarity sum is below the physical threshold. Consider for example the kaon electromagnetic form factor $F_K(t)$, defined by the matrix element

$$\langle K^+(p')K^-(p); \text{out}|J_\mu^{\text{elm}}(0)|0\rangle = (p'-p)_\mu F_K(t), \quad t=(p+p')^2. \quad (1.67)$$

By applying the LSZ formalism as for the pion electromagnetic form factor, one can show that $F_K(t)$ has a cut for $t \geq 4m_\pi^2$, with a discontinuity given by unitarity as

$$\text{Im} F_K(t+i\varepsilon) = \rho(t)T_1(t)F_\pi^*(t)\theta(t-4m_\pi^2) + \cdots \quad (1.68)$$

where $T_1(t)$ is the P wave of the amplitude of the process $K^+K^- \to \pi^+\pi^-$ continued in the unphysical region, at values of t below the physical threshold $t = 4m_K^2$. Such situations arise also in the case of scattering amplitudes, for instance for $\pi\bar{\pi} \to N\bar{N}$, which is the crossed channel of the pion nucleon scattering, $\pi N \to \pi N$. The validity of the unitarity relation when the absorptive part extends below the physical threshold was discussed for the first time by Mandelstam [25].

A more subtle complication appears when a form factor or a scattering amplitude has branch points not determined by unitarity. Such branch points are known as "anomalous thresholds", and appear for special values of the external masses and of the masses of the particles entering the unitarity sum. The standard technique of treating anomalous thresholds [25–27] is to start from a mass configurations where only normal thresholds appear and to perform an analytic continuation in the external mass parameters as to smoothly connect a normal threshold to an anomalous one. What happens actually is that a singularity initially located on a higher Riemann sheet "creeps" onto the physical sheet, dragging with it and deforming the unitarity cut towards lower values of energies. The anomalous thresholds have been found also in Feynman diagrams, the triangular graphs being typical cases for this phenomenon. The discontinuity across an anomalous cut cannot be calculated from unitarity and requires additional information.

There have been discussions in the 1970s on the connection between anomalous thresholds and compositeness. A nice example is the electromagnetic form factor of the deuteron, where the lowest anomalous threshold is much below the lowest normal unitarity threshold $t = 4m_\pi^2$, and is due to the fact that the deuteron is a weakly bound state of two nucleons (for a detailed discussion of the deuteron form factor see [7]).

1.3.2 $\pi\omega$ Electromagnetic Form Factor

Besides the electromagnetic form factors of the stable hadrons, like the pion, the nucleon or the kaon, of much recent interest are the "transition form factors" which parametrize the contribution of the multiparticle states in physical processes. One example is the $\pi\omega$ electromagnetic form factor $f_{\omega\pi}(t)$ defined as

$$\langle \omega(p_a, \lambda)\pi(p_b)|J_\mu^{(1)}(0)|0\rangle = i\varepsilon_{\mu\tau\rho\sigma}\varepsilon^{\tau*}(p_a, \lambda)p_b^\rho q^\sigma f_{\omega\pi}(t), \quad (1.69)$$

where $J_\mu^{(1)}$ is the isovector part of the electromagnetic current, λ denotes the ω polarization, $q = p_a + p_b$ and $t = q^2$. The form factor $f_{\omega\pi}(t)$ has dimension GeV^{-1}.

Unitarity implies that $f_{\omega\pi}(t)$ has a cut along the real axis for $t \geq 4m_\pi^2$ and its discontinuity across the cut in the elastic approximation is given by

$$\text{disc}\, f_{\omega\pi}(t) \equiv f_{\omega\pi}(t + i\varepsilon) - f_{\omega\pi}(t - i\varepsilon) = \frac{i\, q_{\pi\pi}^3(t)}{6\pi \sqrt{t}} f_1(t)(F_\pi^V(t))^*, \qquad (1.70)$$

where $q_{\pi\pi}(t) = \sqrt{t/4 - m_\pi^2}$, $F_\pi^V(t)$ is the pion electromagnetic form factor, and $f_1(t)$ the P partial-wave amplitude of the scattering process

$$\pi^+(q_1)\, \pi^-(q_2) \to \omega(p_a, \lambda)\, \pi^0(p_b). \qquad (1.71)$$

The scattering process is physical for $t \geq (m_\omega + m_\pi)^2$. In the region $4m_\pi^2 \leq t < (m_\omega - m_\pi)^2$, where the decay $\omega \to \pi^+\pi^-\pi^0$ is allowed, $f_1(t)$ is the P-wave projection of the decay amplitude, and the region $(m_\omega - m_\pi)^2 \leq t < (m_\omega + m_\pi)^2$ is unphysical.

We recall that he expression (1.70) of the discontinuity is valid in the region $4m_\pi^2 \leq t < (m_\omega + m_\pi)^2$, since above the $\omega\pi$ threshold other intermediate states contribute in the unitarity sum.

As discussed above, the phase of the form factor in the elastic region is equal to the $\pi\pi$ P-wave phase shift δ_1^1. On the other hand, contrary to naive expectations, the phase of the partial wave $f_1(t)$ of the process (1.71), does not coincide with the $\pi\pi$ P-wave phase shift. More precisely, in the projection of the total amplitude onto the P partial wave, the kinematical variables of the process (1.71) reach regions where the decay

$$\omega(p_a, \lambda) \to \pi^+(q_1)\, \pi^-(q_2)\, \pi^0(-p_b) \qquad (1.72)$$

is allowed. Therefore, rescattering between the three final pions is possible, contributing to the phase of $f_1(t)$. As a consequence, the phases of $f_1(t)$ and $(F_\pi^V(t))^*$ do not compensate each other in (1.70), the discontinuity is not purely imaginary and $f_{\omega\pi}(t)$ is not an analytic function of real type (in particular, it is not real for $t < 4m_\pi^2$, below the unitarity threshold).

One might hope to reinstate the Schwarz reflection property by applying the trick of Goldberger and Treiman. However, it turns out that this is not possible. The situation of the $\pi\omega$ form factor is different from that of the pion form factor, due to the fact that ω is not a stable particle (it decays into three pions by a strong interaction). In the LSZ formalism and in the arguments based on PT invariance used above, it was essential that the particles are "asymptotic states", i.e. are stable with respect to the strong interactions, which is not the case for ω. In particular, the invariance under PT operator, valid for one-particle asymptotic states, is no longer valid for ω. We can understand this from the fact that ω can be viewed as a certain subspace of the multiparticle Fock space. Thus, we cannot apply an argument similar to that in (1.48), which led in the case of the pion form factor to the conclusion that the function $G(t)$ was real.

Recall also that in the Goldberger and Treiman procedure we used the fact that the "in" and "out" one-particle states coincide. Again, this is no longer valid for ω, which is not rigorously a one-particle state. So, the discontinuity of the $\pi\omega$ form factor across the cut cannot be made purely imaginary. We conclude that $f_{\omega\pi}(t)$ is an analytic function in the t-plane cut for $t \geq 4m_\pi^2$, but it does not satisfy the Schwarz reflection. This feature is encountered also in the case of other transition form factors involving particles unstable with respect to strong interactions.

1.4 Dispersion Relations for Scattering Amplitudes

In this section we briefly review the standard dispersion relations for the scattering amplitudes. We first derive the dispersion relations for $T(s, t)$ at fixed t, discuss the problem of subtractions and the behavior at large energies in Regge theory, and finally write down the dispersion relations in two variables known as Mandelstam representation.

We described in Sect. 1.2 the derivation of analyticity properties for the hadron scattering amplitudes in the frame of axiomatic QFT. In particular, we saw that for fixed t in a certain region (in particular for $t = 0$), the amplitude $T(s, t)$ defined in (1.7) is analytic in the complex s plane cut along the real axis as shown in Fig. 1.3, and satisfies Schwarz reflection property (1.33).

Mathematically, the most straightforward way to exploit analyticity is through Cauchy integral relation. Assume that there are no poles (bound states) on the first Riemann sheet and apply Cauchy relation to the amplitude $T(s, t)$ as function of s at fixed t, taking the contour C as shown in Fig. 1.5, around the cuts along the real axis and closed by a circle of large radius R:

$$T(s, t) = \frac{1}{2\pi i} \oint_C \frac{T(s', t)}{s' - s} \, ds'. \qquad (1.73)$$

Due to the opposite directions, the integrals along the real axis above and below the cuts appear with opposite signs, leading to integrals over the discontinuity (1.32),

Fig. 1.5 Integration contour in the complex s-plane for the proof of dispersion relations for $T(s, t)$ at fixed t

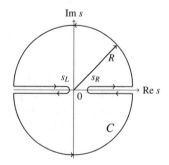

which takes the form (1.34) by Schwarz reflection property. If $|T(s,t)| \to 0$ for $|s| \to \infty$, we can take the limit $R \to \infty$ and neglect the integral along the circle in Fig. 1.5, writing

$$T(s,t) = \frac{1}{\pi} \int_{-\infty}^{s_L} \frac{\operatorname{Im} T(s' + i\varepsilon, t)}{s' - s} ds' + \frac{1}{\pi} \int_{s_R}^{\infty} \frac{\operatorname{Im} T(s' + i\varepsilon, t)}{s' - s} ds'. \tag{1.74}$$

In particular, when s approaches the real axis, by applying the relation (1.29), we write (2.9) as a Hilbert transform, i.e. a relation expressing the real part of the function in terms of its imaginary part. For instance, for real $s > s_R$

$$\operatorname{Re} T(s,t) = \frac{1}{\pi} \int_{-\infty}^{s_L} \frac{\operatorname{Im} T(s' + i\varepsilon, t)}{s' - s} ds' + \frac{P}{\pi} \int_{s_R}^{\infty} \frac{\operatorname{Im} T(s' + i\varepsilon, t)}{s' - s} ds', \tag{1.75}$$

where P denotes the Cauchy principal part. This equality was named "dispersion relation", by analogy with the relation between the real (dispersive) and imaginary (absorptive) parts of the optical index of refraction. By extending the analogy, the imaginary part $\operatorname{Im} T(s,t)$ for $s > s_R$ is said to be the absorptive part of the amplitude in the s-channel, denoted as $A_s(s,t)$, while $\operatorname{Im} T(s,t)$ for $s < s_L$ is the absorptive part $A_u(s,t)$ in the u-channel. By introducing the absorptive parts, the relation (1.74) can be written as

$$T(s,t) = \frac{1}{\pi} \int_{-\infty}^{s_L} \frac{A_u(s' + i\varepsilon, t)}{s' - s} ds' + \frac{1}{\pi} \int_{s_R}^{\infty} \frac{A_s(s' + i\varepsilon, t)}{s' - s} ds'. \tag{1.76}$$

This relation can be easily generalized to the case when the asymptotic behavior of $T(s,t)$ is such that the integrals in (1.74) do not converge. A simple way to do this is to first write formally (1.76) for a certain $s = s_1$,

$$T(s_1, t) = \frac{1}{\pi} \int_{-\infty}^{s_L} \frac{A_u(s' + i\varepsilon, t)}{s' - s_1} ds' + \frac{1}{\pi} \int_{s_R}^{\infty} \frac{A_s(s' + i\varepsilon, t)}{s' - s_1} ds', \tag{1.77}$$

and then subtract this relation from (1.76), which leads to

$$T(s,t) = T(s_1, t) + \frac{s - s_1}{\pi} \int_{-\infty}^{s_L} \frac{A_u(s', t)}{(s' - s)(s' - s_1)} ds' + \frac{s - s_1}{\pi} \int_{s_R}^{\infty} \frac{A_t(s', t)}{(s' - s)(s' - s_1)} ds', \tag{1.78}$$

where the absorptive parts are evaluated on the upper rims of the cuts. Due to the additional s' in the denominator, the integrals in the representation (1.78) converge faster than those in the original relation (1.76). If the asymptotic behavior of the absorptive part is such that the integrals still do not converge, the procedure can be repeated, writing (1.78) for a certain $s = s_2$ and subtracting the relation thus obtained from (1.78). Due to the above formal rule for the derivation of (1.78), the latter relation is said to be a "subtracted dispersion relation", while (1.76) is an "unsubtracted" dispersion relation.

From (1.78) one can see that for the evaluation of a subtracted dispersion relation one needs, in addition to the knowledge of the absorptive parts, the value of the amplitude at one or more points, which must be known from independent sources. This is the price to pay for the fact that the amplitude grows at infinity.

As mentioned in the previous section, rigorous upper bounds on the total or differential scattering amplitudes at large energies have been obtained in the axiomatic theory [18–20]. More specific results are predicted within Regge theory [28, 29], which is based on the analytic continuation of the partial-wave amplitudes t_l in the complex plane of the angular momentum l. For completeness, we briefly describe below the derivation of the asymptotic behaviour in this approach.

The starting point is the partial-wave expansion in the t-channel (obtained from (1.11) by the permutation of the variables s and t) and write it in an equivalent form by using Sommerfeld–Watson transform, namely

$$T(s, t) = \frac{i}{2} \int_{\mathscr{C}} (2l + 1) t_l(t) P_l(\cos \theta_t) \frac{dl}{\sin \pi l}, \tag{1.79}$$

where the contour, indicated in Fig. 1.6, lies close to and surrounds the positive real axis. We assume further that $t_l(t)$ have poles (or cuts) in the complex l-plane (one such pole is indicated for illustration in Fig. 1.6). The positions $l = \alpha(t)$ and the residues $\beta(t)$ of these poles depend on t. By deforming the integration contour in the sense indicated by the arrows until it becomes parallel to the imaginary axis, we pick up, by applying the residue theorem, the contributions of the poles in the l plane encountered during the deformation. Using the fact that for large s and fixed $t \leq 0$ one has $\cos \theta_t \sim s$ and the approximation $P_l(z) \sim z^l$ valid for large $|z|$, we obtain the asymptotic behavior

$$T(s, t) \sim \sum_k \beta_k(t) s^{\alpha_k(t)}, \quad s \to \infty, \quad t \leq 0, \tag{1.80}$$

where the sum is over the rightmost poles in the complex l plane. We used the fact that the other poles and the integral along the vertical line give a suppressed contribution.

Fig. 1.6 Complex l-plane and integration contour for the Sommerfeld–Watson transform

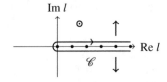

There are theoretical and phenomenological indications that the functions $\alpha_k(t)$ (denoted as Regge trajectories), are approximately linear functions, $\alpha(t) \sim \alpha(0) + \alpha' t$. For $t < 4m^2$, i.e. below the unitarity threshold in the crossed t-channel, the trajectories are real, while for $t \geq 4m^2$ they become complex. Therefore, a pole of the t-channel partial wave $t_l(t)$ can be written as

$$t_l(t) \sim \frac{\beta}{l - \alpha(t)} \sim \frac{\tilde{\beta}}{t - M_R^2 - i\Gamma}, \tag{1.81}$$

which shows that the Regge poles describe the resonances (unstable particles) of complex mass in the crossed channel. Regge theory, originally developed in potential scattering [28], has been widely used in high-energy physics [29].

It is useful to mention one particularly interesting extension of the dispersion relation (1.76), known as Mandelstam representation. The real breakthrough that Mandelstam made was to show how scattering amplitudes could be thought of as functions of more than one complex variable.

Guided by analytic properies that could be established in low order perturbation theory in quantum field theory, and from non-relativistic scattering of a particle off a potential, Mandelstam [30] proposed to apply Cauchy theorem twice and write a double dispersion relation of the form (neglecting poles and possible subtractions):

$$T(s, t, u) = \frac{1}{\pi^2} \iint \frac{\rho_{st}(s', t')}{(s' - s)(t' - t)} \, ds' dt' + \frac{1}{\pi^2} \iint \frac{\rho_{tu}(t', u')}{(t' - t)(u' - u)} \, dt' du'$$
$$+ \frac{1}{\pi^2} \iint \frac{\rho_{us}(s', u')}{(s' - s)(u' - u)} \, du' ds', \tag{1.82}$$

where the integrals are over suitable domains (determined from unitarity) where the real-analytic double spectral functions ρ_{st}, ρ_{tu} and ρ_{us} are nonzero. The representation (1.82) implies that $T(s, t, u)$ is analytic in the topological product of the complex planes s, t and u cut along parts of their real axes. Although a complete proof of the Mandelstam representation (1.82) is not available, it was widely used especially for the dispersive analyses of partial waves for processes of phenomenological interest like πN, $\pi\pi$ or πK elastic scattering. Also, many interesting results concerning the comparison of the axiomatic predictions and those of the Mandelstam representation have been derived [19, 20], in particular for elastic $\pi\pi$ scattering. We emphasize that the analyticity domain implied by Mandelstam representation is much larger than the domains derived from axiomatic field theory.

1.5 Dispersion Theory for Form Factors

In this section we shall discuss the main approach of exploiting analyticity in the case of form factors, which is based on a particular dispersion relation known as Omnès representation.

As we saw in Sect. 1.3, Fermi–Watson theorem (1.59) states that the phase of the pion vector form factor in the elastic region is equal to the phase shift $\delta_1^1(t)$ of the $\pi\pi$ elastic scattering. Similar relations hold also for other form factors. Although the phase is not known above the inelastic threshold, the mathematical problem of constructing an analytic function from the knowledge of its phase along the whole cut presents its own interest, and has been solved by physicists a long time ago [31].

We start from the relation (1.58) and take the logarithm in both sides. Using in addition the Schwarz reflection property (1.60) and recalling that by $F(t)$ in (1.58) we mean $F(t + i\varepsilon)$, while $F^*(t)$ is actually $F(t - i\varepsilon)$, we can write

$$\ln F(t + i\varepsilon) - \ln F(t - i\varepsilon) = 2i\varphi(t), \quad t \geq 4m_\pi^2, \tag{1.83}$$

where the phase $\varphi(t)$ is known along the whole cut, being equal to $\delta_1^1(t)$ for $t < t_{in}$.

In the left-hand side of this relation we recognize the discontinuity across the cut of the function $\ln F(t)$. If this function were analytic in the cut t-plane and decrease at large t, we could construct the function in terms of its discontinuity by applying Cauchy integral relation as in the previous section. However, it is known that the logarithm is singular at the points where $F(t) = 0$. Therefore, before applying Cauchy relation, we must eliminate the zeros of the function in the holomorphy domain. Let us denote by $P(t)$ the polynomial which includes these zeros, i.e. $P(t_k) = 0$ at the points t_k where $F(t)$ vanishes in the complex plane.[5] Defining

$$\Omega(t) = \frac{F(t)}{P(t)}, \tag{1.84}$$

we obtain from (1.83)

$$\ln \Omega(t + i\varepsilon) - \ln \Omega(t - i\varepsilon) = \ln F(t + i\varepsilon) - \ln F(t - i\varepsilon), \tag{1.85}$$

since the polynomial is regular in all the complex plane. Moreover, $\ln \Omega(t)$ is analytic in the cut t plane, since $\Omega(t)$ has no zeros. Therefore, we can obtain $\ln \Omega(t)$ from the Cauchy relation in terms of the discontinuity (1.83). In the most general case, assuming that the phase $\varphi(t)$ approaches asymptotically a nonzero constant $\varphi(\infty)$, the dispersion relation requires a subtraction. Taking for convenience the subtraction point t_1 at the origin, $t_1 = 0$, we obtain

$$\ln \Omega(t) = \ln \Omega(0) + \frac{t}{\pi} \int_{4m_\pi^2}^{\infty} \frac{\varphi(t')}{t'(t' - t)} dt'. \tag{1.86}$$

We can take without loss of generality $\Omega(0) = 1$, since this constant can be absorbed in the polynomial $P(t)$. Then (1.86) leads to the representation

---

[5]The zeros on the cut can be accounted for by a suitable variation by $\pm\pi$ of the phase.

$$\Omega(t) = \exp\left[\frac{t}{\pi}\int_{4m_\pi^2}^\infty \frac{\varphi(t')}{t'(t'-t)}dt'\right]. \tag{1.87}$$

This function is known as an Omnès function [16, 31]. It is analytic and without zeros in the cut plane, is normalized by $\Omega(0) = 1$ and satisfies the condition arg $\Omega(t + i\varepsilon) = \varphi(t)$ for $t \geq 4m_\pi^2$. It is easy to derive also the asymptotic behavior [7]

$$\Omega(t) \sim t^{-\frac{\varphi(\infty)}{\pi}}, \quad |t| \to \infty. \tag{1.88}$$

Going back to the form factor $F(t)$, we write it in the most general form as

$$F(t) = P(t)\,\Omega(t), \tag{1.89}$$

where the polynomial is arbitrary, the only condition being $P(0) = 1$ which follows from the normalization (1.61). We note also that

$$F(t) \sim t^{n-\frac{\varphi(\infty)}{\pi}}, \quad |t| \to \infty, \tag{1.90}$$

where n is the number of zeros.

The conclusion of this subsection is that an analytic function can be reconstructed from its phase on the cut, provided the positions of the zeros in the complex plane are known. In practical applications, however, the phase is not always known along the whole cut. What complementary information may be available and how to treat in an optimal way such situations will be the subject of the forthcoming chapters of this book.

1.6 Instability of Analytic Extrapolation

In the previous sections we saw that analyticity, combined with unitarity and crossing symmetry, played a great role in the physics of strong interactions in the 1960–1970s. Analytic continuation was used to connect the physical region of one process to the physical regions of other processes, the so-called crossed channels. The power of analyticity is expressed by an exact mathematical property, namely the uniqueness of analytic continuation. This means, in particular, that if two functions analytic in a certain domain of the complex plane coincide along a contour inside the analyticity domain, they are identical in the whole domain. This is a good facet of analyticity, its veritable "splendour". However, there is also another, more dangerous facet, which is the fact that analytic continuation is an unstable problem ("ill-posed" in the Hadamard sense).

In mathematics, an ill-posed problem is one which does not meet the Hadamard criteria for being well-posed, which require that the problem has a unique solution that depends continuously on the parameters or input data. For analytic continuation, the criterion of a unique solution is satisfied, however the process does not depend continuously on the initial data. This means that two analytic functions very close along a range Γ in the complex plane may differ arbitrarily much outside Γ. Since in applications we often encounter situations when only an approximation of the exact function is available, this feature of analytic continuation is expected to produce real troubles. In particle physics, first discussions of this aspect of analytic continuation with specific examples have been given in [32–34]. Further works have been devoted to this problem (see [35–39] and references therein).

The ill-posed problems in the Hadamard sense require a regularization to make them stable and susceptible of being solved by a stable, sometimes numerical, algorithm. Tikhonov regularization [40] is the most commonly used method of regularization of ill-posed problems. Loosely speaking, this method amounts to look for solutions of the problem belonging to a more restricted space, usually a compact set.[6]

In order to see what type of conditions can act as a regularization for analytic continuation, we shall present below an example discussed in [38]. Namely, let $f_1(z)$ and $f_2(z)$ be two analytic functions which are very close along a part of the boundary of analyticity domain. Then it is possible to show that their difference can be arbitrarily large at points outside the original domain.

For simplicity, denote $f(z) = f_1(z) - f_2(z)$ and consider that the analyticity domain is the unit disk $|z| < 1$. We further assume that the function $f(z)$ is of real type, i.e. it satisfies the condition $f(z^*) = f^*(z)$. Let us consider the following problem.

Problem Let $\mathscr{F}$ be the class of functions $f(z)$ real analytic in $|z| < 1$, which satisfy the boundary condition

$$|f(\zeta)| \leq \varepsilon, \quad \zeta \in \Gamma_1, \tag{1.91}$$

where Γ_1 is a part of the boundary $|z| = 1$ and $\varepsilon > 0$ is a given number, which may be very small.

Let z_0 a point in the unit disk, outside Γ_1, and δ_0 an arbitrary given number. Then one can find a function $f \in \mathscr{F}$ which satisfies the condition

$$f(z_0) = \delta_0. \tag{1.92}$$

The existence of a function $f(z)$ which simultaneously fulfills the requirements (1.91) and (1.92) is not a priori obvious. Intuitively, one may think that, if ε is small, the function $f(z)$ must be small also at points z_0 not very far from Γ_1, and this would make impossible to satisfy (1.92), for an a priori fixed δ_0. However, it turns out that this intuitive expectation is not met. We shall solve the above problem in Sect. 4.8, using mathematical techniques which we shall present in Chap. 3. Recalling the definition of $f(z)$ as the difference $f_1(z) - f_2(z)$, we can formulate the result by

---

[6]A compact topological space is the generalization of the sets closed and bounded in the Euclidean spaces. Criteria for compactness for functional spaces have been formulated, for instance by Arzelà–Ascoli theorem [41].

saying that the functions f_1 and f_2, arbitrarily close to each other along the contour Γ_1 may differ by an arbitrary amount at a point z_0 outside Γ_1. This is precisely the instability of analytic continuation.

From the explicit construction presented in Sect. 4.8 it will be clear what are the situations when the problem does not admit a solution: we will see that this happens if the functions in the class $\mathscr{F}$ satisfy a suitable boundedness condition on the remaining part Γ_2 of the boundary. This particular example shows that in order to perform a stable analytic extrapolation, one must find a stabilizing condition and restrict the class of admissible functions to a compact set.

In practice, actually, very often the opposite situation occurs, i.e. a very narrow class of functions (usually, a specific parametrization) is adopted for making the analytic continuation. While one might think of controlling the extrapolation, this can lead to a serious underestimate of the extrapolation error, since the results can be very different for a different parametrization. Therefore, when performing a reliable analytic extrapolation, balance has to be kept between two opposite requirements: on the one hand, narrow the admissible set sufficiently for achieving stability, on the other hand, do not cut it too drastically (by, for instance, a specific, biased representation, that might not include the physical amplitude itself).

A typical example is the use of Breit–Wigner parametrizations for detecting broad resonances, i.e. poles on the higher Riemann sheets far from the physical region. The pitfalls of the analytic continuation showed themselves in a striking way in the detection of the famous σ or $f_0(500)$ resonance, the lowest scalar isoscalar $\pi\pi$ resonance: the large errors on its mass and width and the big disparities between different predictions led even to the exclusion of this particle from the PDG tables for a period. More sophisticated dispersion relations for $\pi\pi$ amplitudes expressed as a set of exact integral equations known as Roy equations [42], and the increased accuracy of data restored eventually its status as an accepted physical resonance. Other examples of stable analytic continuation, where the available physical information is exploited with suitable mathematical tools, will be encountered in this book.

References

1. R.J. Eden, P.V. Landshoff, D.I. Olive, J.C. Polkinghorne, *The Analytic S-Matrix* (Cambridge University Press, Cambridge, 1966)
2. L.D. Landau, Nucl. Phys. B **13**, 181 (1959)
3. R.E. Cutkosky, J. Math. Phys. **1**, 429 (1960)
4. A. Wightman, Phys. Rev. **101**, 860 (1956)
5. H. Lehmann, K. Symanzik, W. Zimmermann, Nuovo Cim. **1**, 205 (1956); **2**, 425 (1957)
6. G. Källen, A.S. Wightman, Mat-fyz. Skrifft 1, Nr. 6 (1958)
7. G. Barton, *Introduction to Dispersion Techniques in Field Theory* (Benjamin, New York, 1965)
8. R. Oehme, Nuovo Cim. **4**, 1316 (1956)
9. K. Symanzik, Phys. Rev. **105**, 743 (1957)
10. M. Gell-Mann, M.L. Goldberger, W.E. Thirring, Phys. Rev. **95**, 1612 (1954)
11. M.L. Goldberger, Y. Nambu, R. Oehme, Ann. Phys. (New York) **2**, 226 (1956)
12. G.F. Chew, M.L. Goldberger, F.E. Low, Y. Nambu, Phys. Rev. **106**, 1337 (1957)

13. H.J. Bremerman, R. Oehme, J.G. Taylor, Phys. Rev. **100**, 2178 (1958)
14. N.N. Bogoliubov, B.V. Medvedev, M.K. Polivanov, *Voprossy teorii dispersionykh sootnoshenii (Moscow, 1958) [Problems in the Theory of Dispersion Relations]* (Institute for Advanced Study Press, Princeton, 1958)
15. H. Lehmann, Nuovo Cim. **10**, 579 (1958); Suppl. Nuovo Cim. **14**, 1 (1959)
16. N.I. Muskhelishvili, *Singular Integral Equations* (Noordhoff, Groningen, 1953)
17. A. Martin, Nuovo Cim. **42**, 930 (1966)
18. M. Froissart, Phys. Rev. **123**, 1053 (1961)
19. A. Martin, Phys. Rev. **129**, 1432 (1963)
20. A. Martin, F. Cheung, *Analyticity Properties and Bounds of the Scattering Amplitudes* (Gordon and Beach, New York, 1970)
21. C. Lopez, G. Mennessier, Phys. Lett. **58B**, 437 (1975)
22. E. Fermi, Nuovo Cim. **2S1**, 17 (1955) [Riv. Nuovo Cim. **31**, 1 (2008)]
23. K.M. Watson, Phys. Rev. **95**, 228 (1954)
24. M.L. Goldberger, S.B. Treiman, Phys. Rev. **110**, 1178 (1958); Phys. Rev. **111**, 254 (1958)
25. S. Mandelstam, Phys. Rev. Lett. **4**, 84 (1960)
26. R. Oehme, Phys. Rev. **111**, 143 (1958); Nuovo Cim. **13**, 778 (1959); Phys. Rev. **117**, 1151 (1960)
27. Y. Nambu, Nuovo Cim. **9**, 610 (1958)
28. T. Regge, Nuovo Cim. **14**, 951 (1959)
29. P.D.B. Collins, *Introduction to Regge Theory and High Energy Physics* (Cambridge University Press, Cambridge, 1977)
30. S. Mandelstam, Phys. Rev. **112**, 1344 (1958)
31. R. Omnès, Nuovo Cim. **8**, 316 (1958)
32. R.E. Cutkosky, Ann. Phys. **54**, 350 (1969)
33. R.E. Cutkosky, B.B. Deo, Phys. Rev. D **1**, 2547 (1970)
34. S. Ciulli, Nuovo Cim. A **61**, 787 (1969); Nuovo Cim. A **62**, 301 (1969)
35. I. Caprini, S. Ciulli, A. Pomponiu, I. Sabba-Stefanescu, Phys. Rev. D **5**, 1658 (1972)
36. S. Ciulli, Lect. Notes Phys. **17**, 70 (1973)
37. S. Ciulli, G. Nenciu, J. Math. Phys. **14**, 1675 (1973)
38. S. Ciulli, C. Pomponiu, I. Sabba Stefanescu, Phys. Rep. **17**, 133 (1975)
39. I. Caprini, M. Ciulli, S. Ciulli, C. Pomponiu, I. Sabba-Stefanescu, M. Sararu, Comput. Phys. Commun. **18**, 305 (1979)
40. A.N. Tikhonov, Dokl. Akad. Nauk. **39**, 195 (1943)
41. W. Rudin, *Real and Complex Analysis* (McGraw-Hill, New York, 1966)
42. S.M. Roy, Phys. Lett. B **36**, 353 (1971)

Chapter 2
Modern Approach to Analyticity

In this chapter we discuss the status of analyticity within the Standard Model of particle physics. First, the proof given by Oehme for the analytic properties of the hadronic amplitudes in quantum chromodynamics (QCD) is presented at a qualitative level. We then review the recent progress in modern dispersion theory, in connection with the developments in chiral perturbation theory and lattice QCD, and emphasize the role of analytic continuation for relating the predictions of perturbative QCD to low-energy physical observables. Finally, we introduce functional analysis techniques as alternatives to the standard dispersion relations for implementing in an optimal way the available theoretical or experimental input. Two examples involving hadronic form factors and perturbative QCD are given for illustration.

2.1 Oehme's Proof of Analyticity in QCD

In quantum chromodynamics (QCD), the nonabelian gauge theory [1] which explains the strong interactions in the Standard Model, the elementary fields are the quark and gluon fields, and the hadrons observed in nature are composite states of quarks and gluons. The crucial property of asymptotic freedom, which states that at small distances the quarks and gluons behave like free particles, allows the systematic calculation of the correlation functions by perturbation theory, i.e. by Feynman graphs involving free quarks and gluons. On the other hand, the quarks and gluons cannot exist as free asymptotic states, being "confined" within the hadrons. In QCD, the hadrons are composite states of quarks and gluons, invariant under the SU(3) colour group, i.e. they are "colourless". Only colourless states can be produced and observed as free, asymptotic particles, and only their scattering can be measured experimentally.

Quantum chromodynamics is seemingly simple, and its consequences are straightforward in the domain of hard scattering where perturbation theory applies. However, the existence of hadrons, their properties and their binding into nuclei do not appear in the QCD Lagrangian. They all emerge as a result of its strong coupling. One might

© The Author(s), under exclusive licence to Springer Nature Switzerland AG 2019
I. Caprini, *Functional Analysis and Optimization Methods in Hadron Physics*,
SpringerBriefs in Physics, https://doi.org/10.1007/978-3-030-18948-8_2

ask whether the analyticity properties of the hadronic amplitudes remain the same as those derived in the 1950–1960s, within the framework of the general postulates of relativistic quantum field theory formulated in terms of hadron fields, having in view that the hadrons are no longer elementary particles.

The problem has been investigated by R. Oehme, one of the pioneers of the axiomatic approach mentioned in the previous chapter. In a series of papers [2–4], he has asked precisely this legitimate question: what are the singularities of the hadronic amplitudes in QCD and what dispersion relations can be proved in a rigorous way for them?

The answer to this question is of crucial interest, since analyticity and dispersion relations have been the main tool for investigating and understanding a rich amount of phenomena regarding the scattering of hadrons. Moreover, strong interactions play a major role through radiative corrections also in weak and electromagnetic processes, where again dispersive relations have been the most fruitful tools in many situations. A radical change of the analytic properties in the world of hadrons would affect the whole field of "precision physics" which plays a major role in detecting "new physics" beyond the Standard Model.

As we briefly discussed in Chap. 1, the original derivations of dispersion relations are based on the general postulates of relativistic quantum field theory of hadrons. There is no need to specify the theory in detail. The essential input is relativistic covariance, locality (microscopic causality) and spectral conditions. The mathematical framework is the theory of functions of several complex variables, and the main tool is the edge-of-the-wedge theorem. As mentioned in Chap. 1, hadronic perturbation theory at low orders was also helpful for finding a minimal singularity structure.

Fortunately, it turns out that the original proofs of analyticity and of dispersion relations for hadronic amplitudes and form factors remain applicable even if the hadrons are no longer elementary particles. The framework used in [2–4] is QCD formulation in covariant gauges and BRST algebra. Crucial in the proof is confinement, understood as the exclusion of quarks and gluons from the physical state space, so that they cannot appear as asymptotic "in" or "out" states in hadronic collisions. The proof is based on the algebraic definition of confinement, which is mathematically well-defined and invariant. The spectral condition implies that only hadronic states are relevant and ensure the unitarity of the S-matrix. A subtle point is the construction of the local composite fields for hadrons, defined in terms of the fundamental fields of quarks. With these fields and the corresponding asymptotic fields, it is possible to obtain representations of the scattering amplitudes and vertex functions in terms of the advanced and retarded products of Heisenberg fields by using the LSZ formalism. The Fourier representation of the hadronic amplitudes allows the derivation of analytic properties, which follow from the support of the Fourier integrands, determined from microcausality and spectral conditions. The sophisticated mathematical techniques developed previously for the hadronic physics can then be applied to derive specific analytic properties and dispersion relations. The limitations of proofs are in general related to the restricted ability of exploiting unitarity in order to remove unphysical anomalous thresholds.

The conclusion of the analysis presented in [2–4] is that the hadronic scattering amplitudes and Green functions have only poles and branch points determined by hadronic contributions in the unitarity sums. Neither ordinary nor anomalous thresholds which are directly associated with the underlying quark-gluon structure appear. In particular, the absence of anomalous thresholds contrasts with the anomalous thresholds present in vertex functions of composite particles with observables constituents, which we briefly mentioned at the end of Sect. 1.3.1.

2.2 Modern Dispersion Theory in Hadron Physics

The arguments presented in the previous section validate the dispersion theory for hadrons as a sound theoretical instrument within the Standard Model. Actually, considerable progress in the applications of the standard dispersion relations to low-energy hadron physics has been achieved in the last decades.[1] It was due to a large extent to the theoretical input provided by chiral perturbation theory (χPT) and lattice QCD, as well as to the experimental measurements with improved precision.

In this section we briefly review the role played by χPT and lattice QCD and outline several important results obtained recently in dispersion theory. Finally, using for illustration the amplitude of proton Compton scattering and the pion electromagnetic form factor, we show that the input information may appear in a form which is not suitable for the straightforward application of the standard dispersion relations.

Chiral perturbation theory, the effective field theory which implements the low-energy properties of QCD, is an important source of low-energy input in the dispersion theory. In the original formulation [5, 6], χPT is an expansion in the quark masses (m_q) and p^2, with power counting $m_\pi^2 \sim m_q \sim p^2$. This framework provides symmetry relations among different observables, which are best expressed with the help of a set of linearly independent and universal (i.e. process-independent) low-energy constants (LECs), appearing as coefficients of the polynomial terms in the expansions of these observables. Consider, for instance, the Taylor expansion

$$F_\pi^V(t) = 1 + \frac{1}{6}\langle r^2\rangle_\pi^V t + c_\pi^V t^2 + \cdots \tag{2.1}$$

of the pion electromagnetic form factor defined in (1.45). Expressions of the quadratic charge radius $\langle r^2\rangle_\pi^V$ and the curvature c_π^V in terms of the LECs of χPT to NNLO have been derived and can be found in [7].

Lattice QCD is another approach that yield valuable input in the dispersion theory. Large-scale numerical simulations of lattice QCD allow for the computation of the effects of strong interactions from first principles. In the study of the quark-mass dependence of QCD observables calculated on the lattice, it is common practice to

---

[1]For a recent review see: J. A. Oller, *A Brief Introduction to Dispersion Relations—With Modern Applications*, (SpringerBrief in Physics, 2019).

invoke χPT. A comprehensive review of the current status of lattice results for a variety of physical quantities in low-energy physics is published periodically by the Flavour Lattice Averaging Group (FLAG) (see [8] and references therein).

Both χPT and lattice QCD make predictions on the scattering amplitudes and form factors at limited kinematical ranges, usually at low energies and far from the hadronic branch points. For extending these predictions to larger kinematical regions, an analytic continuation, exploiting the known analytic properties which result from causality and unitarity, is necessary. Thus, dispersion theory is an important tool for extending the predictions of χPT and lattice QCD.

In particular, an impressive progress in the knowledge of pion-pion scattering resulted from the marriage of dispersion theory with χPT. In this case, a set of coupled integral equations for partial waves, the so-called Roy equations [9], have been derived from the dispersion relations which fully implement unitarity and crossing symmetry. By using in the Roy equations the constraints provided by χPT, a very precise determination of the pion-pion scattering lengths a_0^I, i.e. the values of the partial waves $t_0^I(s)$ defined in (1.15) at threshold, was obtained in [10]:

$$a_0^0 = 0.220 \pm 0.005, \quad a_0^2 = -0.0444 \pm 0.0010. \tag{2.2}$$

Using further the scattering lengths as subtraction constants in fixed-t dispersion relations, it was possible to obtain a very good description of the pion-pion amplitude at low energies [10–13]. Moreover, a precise determination of the mass and width of the lowest QCD resonance, known as σ or $f_0(500)$, was obtained by the analytic continuation of the partial wave $t_0^0(s)$ in the complex plane [14, 15], in a framework that avoids the instability of analytic extrapolation discussed at the end of Sect. 1.6.

Many other applications, including coupled-channel treatments, have been performed for a variety of processes. We shall mention below only a few of them. The Roy-Steiner equations, derived from dispersion relations on hyperbolic curves, have been applied to πK scattering [16] and πN scattering [17]. Detailed investigations have been performed for the amplitudes and transition form factors relevant for the hadronic contribution to muon $g - 2$, like the pion transition form factor $\pi \to \gamma^*\gamma^*$ [18]. Recent dispersive treatments include the challenging and notoriously difficult three-body decays $\eta \to 3\pi$ [19] and $\omega \to 3\pi$ [20].

There are, however, situations, when the information required in a standard dispersion relation is not available, but alternative knowledge can be obtained from theory or experiment. As shown in the Sect. 1.4, a standard dispersion relation, like the fixed-t relation (1.74), requires the knowledge of the imaginary part along the entire cuts, at all s. Consider, for instance, the Compton scattering on protons, i.e. the elastic scattering $\gamma p \to \gamma p$. In this case, there are six invariant amplitudes $T_i(\nu^2, t)$, $i = 1, \ldots, 6$, free of kinematical constraints, which at fixed momentum transfer t are analytic in the complex energy ν^2 plane cut for $\nu^2 \geq \nu_0^2(t)$ [21]. The information available for these amplitudes can be expressed through the relations [22]

$$\operatorname{Im} T_i(\nu^2, t) = \rho_i(\nu^2, t), \quad \nu_0^2 \leq \nu^2 \leq \nu_{in}^2, \quad i = 1, \ldots, 6, \tag{2.3}$$

and

$$\sum_{i,j=1}^{6} P_{ij}(v^2, t) T_i^*(v^2, t) T_j(v^2, t) = \sigma(v^2, t), \quad v_{in}^2 \leq v^2, \qquad (2.4)$$

where v_{in}^2 is an inelastic threshold and the functions P_{ij}, ρ_i and σ are known from theory or experiment. The problem is to exploit in an optimal way the relations (2.3) and (2.4) for computing the amplitude at physical points of interest. It can be shown that the standard dispersion relations based on Cauchy relation are not sufficient in this case, and other techniques must be considered [23, 24].

For the form factors, we recall that the standard approach is the representation (1.89), involving the product of a polynomial with the Omnès function (1.87). However, as mentioned at the end of Sect. 1.4, in practical application the phase is not known along the whole cut. For instance, for the pion electromagnetic form factor, after removing an isospin-violating contribution due to the ω resonance, Fermi–Watson theorem

$$\text{Arg}[F_\pi^V(t + i\varepsilon)] = \delta_1^1(t), \qquad t_+ \leq t \leq t_{in}, \qquad (2.5)$$

relates the form factor phase to the phase-shift $\delta_1^1(t)$ of the P-wave of $\pi\pi$ elastic scattering only below the first inelastic threshold t_{in} produced by the $\omega\pi$ state. Above the inelastic threshold, the phase of $F_\pi^V(t)$ is not known.

In order to avoid model-dependent assumptions on the phase, one can use instead the phenomenological and theoretical information available on the modulus $|F_\pi^V(t)|$. In a conservative approach, this information can be expressed as [25]

$$\frac{1}{\pi} \int_{t_{in}}^{\infty} w(t) |F_\pi^V(t)|^2 \, dt \leq 1, \qquad (2.6)$$

where $w(t) > 0$ is a suitable positive-definite weight, for which the integral converges.

The conditions (2.5) and (2.6) define a non-standard problem in dispersion theory, with mixed phase and modulus input along different parts of the cut. Its proper treatment requires more powerful tools of functional optimization.

2.3 Analytic Continuation in Perturbative QCD

Even if the quarks and gluons do not give rise to poles and branch points in the Green functions and hadronic amplitudes, some evidence of the composite structure of hadrons actually appears. Since perturbative QCD is related to the weak coupling limit of the full theory, this evidence should be seen in regions of momentum space where the effective coupling is small as a consequence of asymptotic freedom. In these regions, perturbative QCD is a reasonable tool for the approximate computation of the Green functions. The connection with the low-energy hadronic observables involves an analytic continuation, which we shall discuss in this section.

We consider for illustration the QCD Adler function $D(s)$, defined as (see for instance [26] and references therein):

$$D(s) = -s \frac{d\Pi(s)}{ds}, \qquad (2.7)$$

in terms of the invariant amplitude $\Pi(s)$ of the current–current correlation tensor

$$\Pi_{\mu\nu}(p) = i \int d^4x \, e^{-ipx} \langle 0|T(J_\mu(x)J_\nu(0))|0\rangle = (g_{\mu\nu}p^2 - p_\mu p_\nu)\Pi(s), \qquad (2.8)$$

where $s = p^2$ and J_μ is the electromagnetic hadronic current. From causality and unitarity, one can prove that $\Pi(s)$ is an analytic function of real type, i.e. it satisfies Schwarz reflection $\Pi^*(s) = \Pi(s^*)$ in the complex s-plane cut along the real axis above the lowest hadronic threshold $s_+ = 4m_\pi^2$. The asymptotic behavior ensures that it can be represented by a once-subtracted dispersion relation, written for convenience as

$$\Pi(s) = \Pi(0) + \frac{s}{\pi} \int_{s_+}^{\infty} \frac{\mathrm{Im}\,\Pi(s' + i\varepsilon)}{s'(s' - s)} ds'. \qquad (2.9)$$

The spectral function of this dispersion relation is related by unitarity (the optical theorem) to an observable quantity, namely the cross section for e^+e^- annihilation to hadrons:

$$\mathrm{Im}\,\Pi(s + i\varepsilon) = \frac{1}{12\pi} R(s) \qquad (2.10)$$

where

$$R(s) \equiv \frac{\sigma_{e^+e^- \to \mathrm{hadrons}_{(\gamma)}}(s)}{\sigma_{e^+e^- \to \mu^+\mu^-}(s)}. \qquad (2.11)$$

Alternatively, up to isospin corrections, $\mathrm{Im}\,\Pi(s)$ can be related to the cross section of the hadronic decay of the τ lepton. Namely, the τ hadronic decay width, normalized to the leptonic one, is defined as

$$R_\tau = \frac{\Gamma(\tau \to \nu_\tau + \mathrm{hadrons}(\gamma))}{\Gamma(\tau \to \nu_\tau e \bar{\nu}_e)} = \int_0^{m_\tau^2} ds \, \frac{dR_\tau(s)}{ds}, \qquad (2.12)$$

where

$$\frac{dR_\tau(s)}{ds} = \frac{3(1 + \delta_{\mathrm{EW}})}{\pi m_\tau^2} \left(1 - \frac{s}{m_\tau^2}\right)^2 \left(1 + \frac{2s}{m_\tau^2}\right) \mathrm{Im}\,\Pi(s + i\varepsilon), \qquad (2.13)$$

the factor δ_{EW} accounting for electroweak radiative corrections.

At sufficiently large spacelike momenta $s = -Q^2 < 0$, $D(s)$ can be calculated starting from the gluon and quark degrees of freedom, using QCD perturbation theory. In particular, the function $D(s)$ is renormalization-group invariant and ultraviolet finite, and its perturbative renormalization-group improved expansion reads[2]

$$D_{\text{pert}}(s) = 1 + \sum_{n \geq 1} c_{n,1} \left(\frac{\alpha_s(-s)}{\pi} \right)^n, \tag{2.14}$$

where $\alpha_s(\mu^2)$ is the renormalized strong coupling at the scale μ^2. We recall that at one loop the coupling has the expression

$$\alpha_s(-s) = \frac{1}{\beta_0 \log(-s/\Lambda^2)}, \tag{2.15}$$

where Λ is the QCD parameter in a definite renormalization scheme and $\beta_0 = (33 - 2n_f)/12$ is the first coefficient of the β function which governs the scale-dependence of α_s. In the applications considered in this book, we take the number of flavours $n_f = 3$, so that $\beta_0 = 9/4$.

The connection between the QCD perturbation theory and the experimentally accessible spectral functions at low and moderate energies is not trivial. At low energies, the spectral functions exhibit hadronic branch points and resonances, which can not be reproduced when the perturbative expansions like (2.14) are evaluated on the Minkowski axis. In spite of this difference, a complete description in terms of Lagrangian and physical degrees of freedom should be equivalent, a notion which is referred to as "quark-hadron duality". A first practical implementation of this notion was conceived by Poggio, Quinn and Weinberg [27], who argued that QCD perturbation theory can be used to calculate physical mass-shell processes, provided that both the data and the theory are smeared over a suitable energy range. For instance, a suitable smearing of the observable $R(s)$ defined in (2.11) is given by

$$\bar{R}(s, \Delta) = \frac{\Delta}{\pi} \int_0^{\infty} \frac{R(s')}{(s' - s)^2 + \Delta^2} ds', \tag{2.16}$$

for a certain $\Delta > 0$. Starting from the dispersion relation (2.9), it is easy to check that (2.16) is equivalent to

$$\bar{R}(s, \Delta) = \frac{1}{2i} \left[\Pi(s + i\Delta) - \Pi(s - i\Delta) \right], \tag{2.17}$$

involving the polarization amplitude at complex momenta outside the physical region. Since $\Delta \neq 0$ provides a natural infrared cut-off, perturbation theory is expected to be valid for $\bar{R}(s, \Delta)$, justifying the smearing prescription proposed in [27].

---

[2]The expansion (2.14) is actually divergent, the coefficients $c_{n,1}$ growing factorially as $n!$ at large n. We shall discuss in more detail this aspect in Chap. 6.

Fig. 2.1 Integration contour C in the complex s-plane, used in the Cauchy integral (2.19)

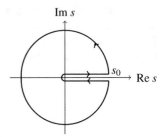

The concept of "quark-hadron duality" was developed along this line and became more precise after its modern implementation [28–30] in the framework of Wilson's operator product expansion (OPE). In OPE, the purely perturbative expansion (2.14) is supplemented by a series of power corrections [31]

$$D_{\text{OPE}}(s) = D_{\text{pert}}(s) + \sum_{k\geq 1} \frac{d_k(s)\langle O_k \rangle}{(-s)^k}, \qquad (2.18)$$

where $\langle O_k \rangle$ are vacuum matrix elements (condensates) of nonperturbative origin and the coefficients $d_k(s)$ depend logarithmically on s.

Quark-hadron duality assumes then that the description (2.18) in the frame of OPE, valid away from the Minkowski axis, can be analytically continued to match with the description in terms of hadrons, which live on the Minkowski axis.

In practical applications, a standard way to implement this idea is based on the Cauchy integral relation

$$\frac{1}{2\pi i} \oint_C \phi(s)\Pi(s)\,ds = 0, \qquad (2.19)$$

where C is the closed contour shown in Fig. 2.1 and $\phi(s)$ is an arbitrary function holomorphic inside the circle $|s| = s_0$. Using the analytic properties of $\Pi(s)$, the relation (2.19) can be written alternatively as

$$\frac{1}{\pi} \int_0^{s_0} \phi(s)\text{Im}\,\Pi(s + i\varepsilon)ds = -\frac{1}{2\pi i} \oint_{|s|=s_0} \phi(s)\Pi(s)\,ds. \qquad (2.20)$$

In phenomenological applications, this relation is exploited by using in the left side the imaginary part $\text{Im}\,\Pi(s + i\varepsilon)$ expressed in terms of observables like (2.10) or (2.13), known for $s \leq s_0$ from experimental measurements, and in the right side the OPE expansion $\Pi_{\text{OPE}}(s)$ from (2.18).

The relation (2.20) is an example of the so-called "finite-energy sum rules" (FESR), extensively used for connecting perturbative QCD to low-energy hadronic physics. Various versions have been proposed (moment sum rules, Borel sum rules) depending on the choice of the holomorphic function $\phi(s)$.

Actually, the input used in the evaluation of the FESR (2.20) is not exact: the data on the spectral function $\text{Im}\,\Pi$ in the l.h.s are affected by experimental errors, and the theoretical expression Π_{OPE} used along the circle is a truncated expansion with an unknown truncation error. Moreover, as it became clear in recent years, additional terms, which violate quark-hadron duality,[3] should be added to $\Pi_{OPE}(s)$. According to the present knowledge, these terms are nonvanishing only in the right half $\mathrm{Re}\, s > 0$ of the s plane, being large in the vicinity of the timelike axis. It turns out that special techniques of complex and functional analysis, more powerful than the simple Cauchy relation, are suitable for dealing with such situations. We shall discuss this alternative approach in Sect. 2.4.2.

2.4 From Dispersion Relations to Optimization Problems

In the previous two sections we argued that there are situations when the input available from the theoretical knowledge or experiment does not allow the application of the standard dispersion relations. We shall show now how such situations can be handled, by expressing the input information in the form of a set of constraints in an extremal problem over a suitable class of functions. We shall illustrate the approach with an example from hadron physics and an example from perturbative QCD.

2.4.1 Extremal Problem for a Hadronic Form Factor

In Sect. 2.2, we argued that Fermi–Watson theorem and the experimental information available on the pion electromagnetic form factor can be expressed by the relations (2.5) and (2.6). Mathematically, these relations represent a mixed problem, where the phase is known along a part of the cut and information on the modulus is available along the remaining part of the cut. A simple, compact formula, similar to the Cauchy relation, allowing to reconstruct the analytic functions from the above relations does not exist. Moreover, the conditions do not specify uniquely the function: there is a whole class of admissible analytic functions consistent with them. Using also the normalization $F_\pi^V(0) = 1$ satisfied by the form factor, we can formulate on this admissible class the following extremal problem:

Problem Denote by $\mathscr{F}$ the class of functions $F(t)$ real analytic in the complex plane cut for $t \geq t_+$, which satisfy the conditions

$$F(0) = 1,$$
$$\mathrm{Arg}[F(t + i\varepsilon)] = \delta(t), \qquad t_+ \leq t \leq t_{in},$$
$$\frac{1}{\pi} \int_{t_{in}}^{\infty} w(t)\,|F(t)|^2\,dt \leq 1, \qquad (2.21)$$

---

[3]We shall discuss the origin and the expression of the duality-violating terms in Sect. 6.1.

where the phase $\delta(t)$ and the suitably-normalized weight $w(t) \geq 0$ are known. Then consider the following extremal problems: find

$$F'_{\text{min}} = \min_{F \in \mathscr{F}} F'(0), \qquad F'_{\text{max}} = \max_{F \in \mathscr{F}} F'(0), \qquad (2.22)$$

and

$$F_{\text{min}} = \min_{F \in \mathscr{F}} |F(t)|, \qquad F_{\text{max}} = \max_{F \in \mathscr{F}} |F(t)|, \qquad (2.23)$$

where $t < t_{in}$ is an arbitrary point taken for convenience on the real axis.

As we shall see in Sect. 4.4.4, the above problem can be solved exactly, yielding optimal values for the bounds in terms of the input conditions given in Eq. (2.21). The conditions adopted ensure the stability of the analytic continuation from the boundary to points inside the analyticity domain, so that the upper and lower bounds on $|F(t)|$ and the derivative $F'(0)$ will be finite. The conditions (2.21) can be generalized by including as input also the values of the form factor at several points t_n inside the analyticity domain or on the cut, below t_{in}. The physical implications of the solution will be discussed in Sect. 5.4.

2.4.2   Functional Analysis Based Approach to Perturbative QCD

The approach based on functional analysis turns out also to be a natural alternative to the standard FESR for implementing quark-hadron duality, i.e. for connecting perturbative QCD to low-energy observables. We illustrate the above statement with the polarization function $\Pi(s)$ considered in Sect. 2.3.

In the derivation of FESR, it is assumed that along the circle in the complex plane in Fig. 2.1 one can replace the exact function $\Pi(s)$ by its approximant $\Pi_{\text{OPE}}(s)$. However, the analyticity properties in the complex momentum plane of the QCD correlation functions calculated with the OPE are different from those of the exact functions. As discussed in Sect. 2.1, the exact Green functions have poles and branch points generated by physical hadron states, while in perturbation theory the branch points are produced by free quarks and gluons. Other, more complicated singularities make their appearance if one goes beyond simple perturbation theory: the renormalization-group improved expansions exhibit the unphysical singularity (Landau pole) at $s = -\Lambda^2$ present in the coupling (2.15), and the power corrections in OPE generate poles at $s = 0$. Moreover, partial resummations of particular terms in the perturbative expansion, like for instance the renormalon chains, introduce other singularities [32].

Therefore, it is necessary to distinguish between the exact Green function, which has the correct analyticity properties, and for which some incomplete information is available from experiment, and the approximant provided by theory, which has different analytic properties. The exact function $\Pi(s)$ cannot coincide with $\Pi_{\text{OPE}}(s)$, even after suitable resummations of the perturbative part.

In order to measure this distinction in a quantitative way, we shall evaluate the maximum value of the difference between the exact function and the truncated OPE along a contour Γ in the complex s plane. In particular, if Γ is the circle $s = s_0 e^{i\theta}$ shown in Fig. 2.1, we consider the difference

$$\delta_\infty(\Pi) = \sup_{\theta \in [0,2\pi]} |\Pi(s_0 e^{i\theta}) - \Pi_{\text{OPE}}(s_0 e^{i\theta})|, \qquad (2.24)$$

where $0 \leq \theta \leq 2\pi$ denotes the angle of the ray in the complex plane (the upper and lower edges of the unitarity cut corresponding to $\theta = 0$ and $\theta = 2\pi$, respectively). In the terminology which will be introduced in Chap. 3, the r.h.s. of this relation represents the L^∞ norm of the difference $\Pi - \Pi_{\text{OPE}}$ on the interval $[0, 2\pi]$. This was indicated by the subscript on the quantity $\delta(\Pi)$ in the l.h.s.

Although the exact function $\Pi(s)$ is not known, some information on it is available, namely its imaginary part at low energies is known from experiment. Then, let us formulate the following functional extremal problem [33]:

Problem Let us denote by $\mathscr{P}$ the class of functions $\Pi(s)$ real analytic in the complex s-plane cut along the real axis for $s \geq s_+$ such that

$$\text{Im}\, \Pi(s + i\varepsilon) = \sigma(s), \qquad s_+ \leq s \leq s_0, \qquad (2.25)$$

where $\sigma(s) > 0$ is known, and let $\delta_\infty(\Pi)$ be a real functional defined on this class by Eq. (2.24). Then consider the following extremal problem:

$$\delta_\infty = \min_{\Pi \in \mathscr{P}} \delta_\infty(\Pi). \qquad (2.26)$$

As we shall see in Sect. 4.6, the solution $\delta_\infty > 0$ of this functional optimization problem is unique and can be computed by means of a numerical algorithm. This will represent a lower bound on the functional distance between the true function and its approximant predicted by OPE. The extremal problem (2.26) will be used in Sect. 6.2 for detecting the violation of quark-hadron duality from experimental data on the spectral function.

References

1. D.J. Gross, F. Wilczek, Phys. Rev. Lett. **30**, 1343 (1973); H.D. Politzer, Phys. Rev. Lett. **30**, 1346 (1973)
2. R. Oehme, PiN Newslett. **7**, 1 (1992)
3. R. Oehme, Mod. Phys. Lett. A **8**, 1533 (1993)
4. R. Oehme, Int. J. Mod. Phys. A **10**, 1995 (1995)
5. J. Gasser, H. Leutwyler, Nucl. Phys. B **250**, 465 (1985)

6. J. Gasser, H. Leutwyler, Ann. Phys. **158**, 142 (1984)
7. J. Bijnens, G. Colangelo, P. Talavera, JHEP **05**, 014 (1998)
8. S. Aoki et al., Eur. Phys. J. C **77**, 112 (2017)
9. S.M. Roy, Phys. Lett. B **36**, 353 (1971)
10. G. Colangelo, J. Gasser, H. Leutwyler, Nucl. Phys. B **603**, 125 (2001)
11. B. Ananthanarayan, G. Colangelo, J. Gasser, H. Leutwyler, Phys. Rept. **353**, 207 (2001)
12. I. Caprini, G. Colangelo, H. Leutwyler, Eur. Phys. J. C **72**, 1860 (2012)
13. R. García-Martín, R. Kamiński, J.R. Peláez, J. Ruiz de Elvira, F.J. Ynduráin, Phys. Rev. D **83**, 074004 (2011)
14. I. Caprini, G. Colangelo, H. Leutwyler, Phys. Rev. Lett. **96**, 132001 (2006)
15. J.R. Peláez, Phys. Rept. **658**, 1 (2016)
16. P. Büttiker, S. Descotes-Genon, B. Moussallam, Eur. Phys. J. C **33**, 409 (2004)
17. M. Hoferichter, J. Ruiz de Elvira, B. Kubis, Ulf-G. Meissner. Phys. Rept. **625**, 1 (2016)
18. M. Hoferichter, B.-L. Hoid, B. Kubis, S. Leupold, S.P. Schneider, JHEP **10**, 141 (2018)
19. G. Colangelo, S. Lanz, H. Leutwyler, E. Passemar, Eur. Phys. J. C **78**, 947 (2018)
20. F. Niecknig, B. Kubis, S.P. Schneider, Eur. Phys. J. C **72**, 2014 (2012)
21. W.A. Bardeen, W.K. Tung, Phys. Rev. **173**, 1423 (1968); Erratum: Phys. Rev. D **4**, 3229 (1971)
22. I. Guiasu, C. Pomponiu, E.E. Radescu, Ann. Phys. **114**, 296 (1978)
23. I. Caprini, I. Guiasu, E.E. Radescu, Phys. Rev. D **25**, 1808 (1982)
24. I. Caprini, Phys. Rev. D **27**, 1479 (1983)
25. B. Ananthanarayan, I. Caprini, D. Das, Phys. Rev. Lett. **119**, 132002 (2017)
26. M. Beneke, M. Jamin, JHEP **09**, 044 (2008)
27. E.C. Poggio, H.R. Quinn, S. Weinberg, Phys. Rev. D **13**, 1958 (1976)
28. M.A. Shifman, Int. J. Mod. Phys. A **11**, 3195 (1996)
29. B. Blok, M.A. Shifman, D.X. Zhang, Phys. Rev. D **57**, 2691 (1998); Erratum: Phys. Rev. D **59**, 019901 (1999)
30. M.A. Shifman, in *At the Frontier of Particle Physics*, vol. 3 (World Scientific, Singapore, 2001), pp. 1447–1494
31. M. A. Shifman, A. I. Vainshtein, V. I. Zakharov, Nucl. Phys. B **147**, 385–448 (1979)
32. I. Caprini, M. Neubert, JHEP **03**, 007 (1999)
33. I. Caprini, M. Golterman, S. Peris, Phys. Rev. D **90**, 033008 (2014)

Chapter 3
Complex and Functional Analysis Tools

In this chapter we collect some mathematical results of complex and functional analysis, which will help the reader to understand the content of the subsequent chapters. We first discuss the properties of the Hardy classes H^p of analytic functions in the unit disk of the complex plane. Then we consider several extremal problems on these spaces and on more general abstract spaces, emphasizing the important role of the so-called duality relations between an extremal problem on a certain space and another problem in the dual space. The techniques presented in this chapter will be applied in the next chapter to specific optimization problems in hadron physics.

3.1 H^p Spaces

We consider functions $f(z)$ analytic in the unit disk $|z| < 1$ of the complex plane. From physical reasons, we will be interested in functions with a restricted growth near the boundary. A reasonable measure of the growth is given by the quantities [1]

$$M_p(r, f) = \left[\frac{1}{2\pi} \int_0^{2\pi} |f(re^{i\theta})|^p d\theta \right]^{1/p},$$
$$M_\infty(r, f) = \max_{\theta \in [0, 2\pi]} |f(re^{i\theta})|. \tag{3.1}$$

A function $f(z)$ analytic in the unit disk $|z| < 1$ is said to be of Hardy class H^p ($0 < p \le \infty$) if $M_p(r, f)$ remains bounded as $r \to 1$ [1]. It is obvious that H^p are linear spaces and $H^q \subset H^p$ if $0 < p < q \le \infty$. In particular, H^∞ is the class of bounded analytic functions in the unit disk, while H^2 is the class of functions which, if expanded as power series

© The Author(s), under exclusive licence to Springer Nature Switzerland AG 2019
I. Caprini, *Functional Analysis and Optimization Methods in Hadron Physics*,
SpringerBriefs in Physics, https://doi.org/10.1007/978-3-030-18948-8_3

$$f(z) = \sum_{n=0}^{\infty} a_n z^n, \tag{3.2}$$

satisfy the condition

$$\sum_{n=0}^{\infty} |a_n|^2 < \infty. \tag{3.3}$$

The classes H^p and H^q are said to be *dual* if the relation

$$\frac{1}{p} + \frac{1}{q} = 1 \tag{3.4}$$

holds [1]. It follows that H^1 and H^∞ are dual to each other, while H^2 is the dual of itself.

Of interest are the boundary values of the functions $f(re^{i\theta})$ when $r \to 1$. It can be shown [1] that for $f \in H^p$ with $p > 0$, the nontangential limit $f(e^{i\theta})$ exists, $\log |f(e^{i\theta})|$ is integrable on $[0, 2\pi]$ and $f(e^{i\theta}) \in L^p$ on the same interval. This means that

$$\|f\|_p \equiv \left(\frac{1}{2\pi} \int_0^{2\pi} |f(e^{i\theta})|^p d\theta \right)^{1/p} < \infty, \qquad 0 < p < \infty, \tag{3.5}$$

and

$$\|f\|_\infty \equiv \operatorname{ess\,sup}_{\theta \in [0, 2\pi]} |f(e^{i\theta})| < \infty. \tag{3.6}$$

We recall that the essential suppremum is defined as [6]

$$\operatorname{ess\,sup} |f(t)| \equiv \inf_{g(t) = f(t)\,\text{a.e.}} |g(t)|, \tag{3.7}$$

where a.e. means "almost everywhere", i.e. except on a set of zero measure.

For the further developments of the theory, it is important that $f(re^{i\theta})$ always tends to $f(e^{i\theta})$ in the sense of L^p mean. From this remark one can conclude that H^p is a normed linear space for $1 \le p \le \infty$. The norm is defined as the L^p norm of the boundary function. Thus, if $1 \le p < \infty$,

$$\|f\|_p = \lim_{r \to 1} M_p(r, f) = \left[\frac{1}{2\pi} \int_0^{2\pi} |f(e^{i\theta})|^p d\theta \right]^{1/p} \tag{3.8}$$

and

$$\|f\|_\infty = \sup_{|z|<1} |f(z)| = \operatorname{ess\,sup}_{0 \le \theta \le 2\pi} |f(e^{i\theta})|. \tag{3.9}$$

3.2   Inner and Outer Functions, Canonical Factorization Theorem

A known result in complex analysis [1, 2] is that the zeros of an analytic function cannot cluster inside the domain of analyticity unless the function vanishes identically. For functions of class H^p, the growth condition (3.1) imposes a fuhrer restriction on the density of zeros. Namely, if z_n denote the zeros of a function $f(z)$ of class H^p, then it is shown [1] that

$$\sum_n (1 - |z_n|) < \infty. \tag{3.10}$$

Conversely, given a set of complex numbers z_n which satisfy (3.10), one can construct a bounded analytic function whose zeros are precisely z_n. The construction is based on the simple functions

$$\frac{z - z_n}{1 - z_n^* z} \tag{3.11}$$

which vanish at $z = z_n$ and map the unit disk onto itself (in our notations z_n^* is the complex conjugate of z_n). Thus, one can show [1] that, if z_n are a sequence of complex numbers that satisfy (3.10), the infinite product

$$B(z) = \prod_{n=1}^{\infty} \frac{|z_n|}{z_n} \frac{z - z_n}{1 - z_n^* z} \tag{3.12}$$

converges uniformly in each disk $|z| \le R < 1$, vanishes for $z = z_n$, has no other zeros in $|z| < 1$, is bounded by 1 inside the unit disk and is equal to 1 almost everywhere (a. e.) on the boundary:

$$\begin{aligned} |B(z)| &< 1, & |z| &< 1, \\ |B(e^{i\theta})| &= 1, & 0 &\le \theta < 2\pi, \quad \text{a. e.} \end{aligned} \tag{3.13}$$

A function of the form

$$B(z) = z^m \prod \frac{|z_n|}{z_n} \frac{z - z_n}{1 - z_n^* z}, \tag{3.14}$$

where m is a positive integer and z_n satisfy (3.10) is called a *Blaschke product*. The set z_n may be finite or even empty.

Recalling now that for $f \in H^p$ ($p > 0$), the function $\ln |f(e^{i\theta})|$ is integrable on $[0, 2\pi]$, we define the function

$$F(z) = \exp \left[\frac{1}{2\pi} \int_0^{2\pi} \frac{e^{i\theta} + z}{e^{i\theta} - z} \ln |f(e^{i\theta})| d\theta \right]. \tag{3.15}$$

From the definition it follows that $F(z)$ is analytic in $|z| < 1$ and $|F(e^{i\theta})| = |f(e^{i\theta})|$ a. e. on $[0, 2\pi]$.

According to the standard terminology, a function of the form (3.15) is said to be an *outer function* of class H^p, while a function $f(z)$ analytic in $|z| < 1$ having the properties $|f(z) \leq 1$ and $|f(e^{i\theta})| = 1$ a.e. is called an *inner function*. Clearly, the Blaschke product (3.14) is an inner function. It is sometimes necessary to consider also functions of the form

$$S(z) = \exp\left[-\int_0^{2\pi} \frac{e^{i\theta} + z}{e^{i\theta} - z} d\mu(\theta)\right], \tag{3.16}$$

where $\mu(\theta)$ is a bounded nondecreasing singular function, i.e. $\mu'(\theta) = 0$ a.e. From this representation it follows that

$$0 < |S(z)| \leq 1, \quad |S(e^{i\theta})| = 1 \quad \text{a. e.} \tag{3.17}$$

Such a function $S(z)$ is called a *singular inner function*.

We can now formulate the following *canonical factorization theorem*, quoted as Theorem 2.8 in [1]:

Theorem *Every function $f(z) \not\equiv 0$ of class H^p ($p > 0$) has a unique factorization of the form*

$$f(z) = B(z)S(z)F(z), \tag{3.18}$$

where $B(z)$ is a Blaschke product, $S(z)$ is a singular inner function and $F(z)$ is an outer function for the class H^p, defined by (3.15). Conversely, every such product $B(z)S(z)F(z)$ belongs to H^p.

3.3 Factorization of Matrix-Valued Analytic Functions

This section is a short digression from the main subject of this chapter, in which we consider matrix-valued functions $W(z)$ of a complex variable z. The values of these functions are $n \times n$ matrices with complex elements. We will be interested in particular in analytic matrix-valued functions.

We quote first a result in complex analysis [3], which states that every matrix $P(\zeta)$ hermitian and positive-definite a. e. on the boundary $\zeta = e^{i\theta}$ of the unit disk can be factorized as

$$P(\zeta) = N^{\dagger}(\zeta)\, N(\zeta), \tag{3.19}$$

where $N(z)$ is a matrix analytic and invertible (det $N(z) \neq 0$) inside the unit disk $|z| < 1$, unique up to a constant unitary matrix. We recall that an $n \times n$ hermitian matrix P is said to be positive-definite if the scalar $q^* P q$ is strictly positive for every nonzero column vector q of complex numbers (in our case of the form $e^{i\theta_j}$ with

$\theta_j \in [0, 2\pi)$). Here q^* denotes the conjugate transpose of q. We emphasize that this is only an existence theorem, and there are no practical means to construct explicitly $N(z)$ for a general given $P(\zeta)$.

We quote also a representation which extends the factorization (3.18) to matrix-valued functions $W(z)$ analytic in $|z| < 1$. It is called Blaschke–Potapov factorization and expresses the analytic $n \times n$ matrix $W(z)$ as [4]

$$W(z) = B(z) \exp\left[\int_0^{2\pi} \frac{e^{i\theta} + z}{e^{i\theta} - z} \, d\mu(\theta)\right]. \tag{3.20}$$

Here the Blaschke–Potapov factor $B(z)$ is a matrix of the form

$$B(z) = \prod_k V_k \begin{bmatrix} \frac{|z_k|}{z_k} \frac{z - z_k}{1 - z_k^* z} I_{p_k} & O \\ O & I \end{bmatrix} V_k^{-1}, \tag{3.21}$$

where z_k are the zeros of det $W(z)$ in $|z| < 1$, I_p and I are $p \times p$ and $(n - p) \times (n - p)$ unit matrices, respectively, and V_k are unitary numerical matrices. In the second factor of (3.20), which corresponds to the product $S(z)F(z)$ in the scalar factorization (3.18), the measure $d\mu(\theta)$ is a matrix-valued function, containing both a regular and a singular part.

3.4 Description of the Boundary Values

Coming back to the case of scalar analytic functions, is of interest to characterize in more detail the boundary values of the functions of H^p class on the unit circle $|z| = 1$. We mentioned already that the limits $f(e^{i\theta})$ belong to the L^p spaces. Actually, more can be said about these values. Denote by $\mathcal{H}^p$ the set of boundary functions $f(e^{i\theta})$ of functions $f \in H^p$. As usual, two functions are identified if they coincide almost everywhere, so the elements of $\mathcal{H}^p$ are defined as equivalent classes. It is known that $\mathcal{H}^p \subset L^p$, and in fact $\mathcal{H}^p$ is a vector subspace of L^p: it is closed under addition and scalar multiplication. As proved in [1], it is also topologically closed. Namely, if we consider functions of the form

$$\sum_{k=0}^n a_k e^{ik\theta}, \tag{3.22}$$

denoted as *polynomials in $e^{i\theta}$*, one can show that for $0 < p < \infty$, $\mathcal{H}^p$ is the L^p closure of polynomials in $e^{i\theta}$.

This result is not valid for $p = \infty$, since $\mathcal{H}^\infty$ contains functions which do not coincide almost everywhere with continuous functions. One example is the singular inner function $\exp[(z + 1)/(z - 1)]$, whose boundary function is $\exp[-i \cot(\theta/2)]$, $\theta \neq 0$. However, it can be proved that $\mathcal{H}^\infty$ is closed.

The boundary values of the functions of class H^p can be characterized also by the Fourier coefficients of the boundary functions. The *Fourier coefficients* of a function $\varphi \in L^1$ are the numbers

$$c_n = \int_0^{2\pi} e^{-int}\, \varphi(t)dt, \quad n = 0, \pm 1, \pm 2, \ldots \tag{3.23}$$

An important result is that the Taylor coefficients of a function $f \in H^p$ ($1 \le p \le \infty$) coincide with the Fourier coefficients of its boundary function. More precisely, the following theorem, quoted in [1] as Theorem 3.4, is valid:

Theorem *If $f \in H^p$ has the Taylor expansion*

$$f(z) = \sum_{n=0}^{\infty} a_n\, z^n, \tag{3.24}$$

and c_n are the Fourier coefficients of its boundary function

$$c_n = \int_0^{2\pi} e^{-int}\, f(e^{it})dt, \quad n = 0, \pm 1, \pm 2, \ldots \tag{3.25}$$

then $c_n = a_n$ for $n \ge 0$ and $c_n = 0$ for $n < 0$. Furthermore, $\mathscr{H}^p$ ($1 \le p \le \infty$) is exactly the class of L^p functions whose Fourier coefficients vanish for all $n < 0$.

Due to the these properties, $\mathscr{H}^p$ is often identified in the literature to H^p, and H^p is regarded as a subspace of L^p.

We finally mention that every function $f \in H^p$ ($p \ge 1$) can be expressed in $|z| < 1$ as the Cauchy integral of its boundary function. This property is contained in the following theorem, quoted as Theorem 3.6 in [1]:

Theorem *Every function $f \in H^1$ can be expressed as the Cauchy integral of its boundary function*

$$f(z) = \frac{1}{2\pi i} \int_{|\zeta|=1} \frac{f(\zeta)d\zeta}{\zeta - z}, \quad |z| < 1. \tag{3.26}$$

By the differentiation upon z under the integral sign, we obtain the more general relation

$$f^{(k)}(z) = \frac{k!}{2\pi i} \int_{|\zeta|=1} \frac{f(\zeta)d\zeta}{(\zeta - z)^{k+1}}, \quad |z| < 1; \quad k = 0, 1, 2, \ldots \tag{3.27}$$

3.5 H^p as a Banach Space

We recall that a normed linear vector space X is *complete* if every Cauchy sequence of vectors converges to a well defined limit that is within the space X. A complete normed linear vector space is called a *Banach space*. Banach spaces play a central role in functional analysis [5, 6]. In particular, a *Hilbert space* is a Banach space equipped with an inner product which induces the norm. Thus, if we denote the inner product of two vectors x, $y \in X$ by (x, y), we have

$$\|x\| \equiv \sqrt{(x, x)}. \tag{3.28}$$

As discussed above, $\mathcal{H}^p$ is a topologically closed vector subspace of L^p. It follows that for $1 \le p \le \infty$, H^p is a Banach space with norm $\|f\|_p = M_p(1, f)$. For $p < 1$, it can be shown [1] that $\| \ \|_p$ is not a true norm, and the space H^p is not normable. However, one can define a distance by $d(f, g) = \|f - g\|_p$ even for $0 < p < 1$, and show that H^p for $0 < p < 1$ is a complete metric space.

In particular, on the space L^2 one can define the inner product

$$(f, g) \equiv \frac{1}{2\pi} \int_0^{2\pi} f^*(\theta) g(\theta) \, d\theta, \tag{3.29}$$

such that the norm is

$$\|f\|_2 = \sqrt{(f, f)}. \tag{3.30}$$

Therefore, L^2 is a Hilbert space, and this implies that H^2 is a Hilbert space.

The fact that the H^p spaces can be viewed as Banach spaces is advantageous since it allows the application of techniques of functional analysis for solving extremal and interpolation problems on these spaces.

A major objective in the study of the linear space structure of H^p is to represent the continuous linear functionals on H^p. We recall that a linear functional on a vector space X over a field $\mathbb{K}$ is a mapping $\phi : X \to \mathbb{K}$ such that

$$\phi(f + g) = \phi(f) + \phi(g), \qquad \phi(\lambda f) = \lambda \phi(f), \tag{3.31}$$

for all $f, g \in X$ and $\lambda \in \mathbb{K}$. In particular, we shall consider real or complex linear functionals, $\mathbb{K} \equiv \mathbb{R}$ or $\mathbb{K} \equiv \mathbb{C}$.

A known result in functional analysis is that a continuous linear functional on a normed space is bounded [5]. The space of the bounded linear functionals on X is said to form the *dual space*, denoted as X^*. On the space X^* one can define a norm as

$$\|\phi\| = \sup\{|\phi(f)| : f \in X, \|f\| \le 1\}. \tag{3.32}$$

We note that a more symmetric notation of the dual space is sometimes used in the literature [6], where the action of a linear functional $x^* \in X^*$ on the element $x \in X$ is denoted by $\langle x, x^* \rangle$. With this notation, the relation (3.32) is written as

$$\|x^*\| = \sup\{|\langle x, x^* \rangle| : x \in X, \|x\| \le 1\}. \tag{3.33}$$

In what follows we shall use both these notations.

One of the most famous theorems in functional analysis, which plays a crucial role in optimization theory, is Hahn–Banach theorem. There are several formulations, both analytic and geometrical, of this theorem [5]. In Sect. 3.7 we shall briefly discuss the geometrical version of Hahn–Banach theorem and the geometrical approach to optimization problems in vector spaces, following [6]. Here we give for completeness an analytic formulation useful in the present section.

We recall that S is a *subspace* of a linear space X if every vector of the form $\alpha x + \beta y$ is in S whenever x and y are both in S. Here α and β are scalars, assumed here to be real or complex numbers. Moreover, a subspace is closed if every Cauchy sequence in S has a limit in S.

Theorem (Hahn–Banach) *If X is a normed vector space with a linear subspace S (not necessarily closed) and if $\phi : S \to \mathbb{K}$ is a continuous linear functional, then there exists an extension $\psi : X \to \mathbb{K}$ of ϕ which is also continuous and linear and which has the same norm as ϕ, i.e.*

$$\psi(x) = \phi(x), \; x \in S; \quad \|\psi\|_{X^*} = \|\phi\|_{S^*}. \tag{3.34}$$

Hahn–Banach theorem is most profitably viewed as an existence theorem for a minimization problem. Indeed, given a linear functional ϕ on a subspace S of a normed space, it is not difficult to extend it to the whole space. However, an arbitrary extension will in general be unbounded or have a greater norm than that of ϕ on S. The problem is therefore to select the extension of minimum norm. Hahn–Banach theorem guarantees the existence of a minimum-norm extension and says what is the norm of the best extention.

It is useful to quote two results, which can be proved by using Hahn–Banach theorem and play an essential role in the theory of extremal problems. We first introduce some notations. A *coset of X modulo S* is a subset $\xi = x + S$ consisting of all elements of the form $x + y$ where x is a fixed element of X and y ranges over S. Two cosets are either identical or disjoint. The *quotient space X/S* has as its elements all distincts cosets of X modulo S. With the natural definition of addition and scalar multiplication, X/S is a linear space, the zero element being the coset S. Moreover, the norm of a coset $\xi = x + S$ is defined by

$$\|\xi\| = \inf_{y \in S} \|x + y\|. \tag{3.35}$$

This is a genuine norm: indeed, since S is closed, $\|\xi\| = 0$ implies $\xi = S$. Furthermore, under this norm the space X/S is complete, and thus is a Banach space itself.

We recall also that the *annihilator* of the subspace S is the set $S^\perp$ of all linear functionals $\phi \in X^*$ such that $\phi(x) = 0$ for all $x \in S$. It can be verified that $S^\perp$ is a subspace of X^*. Moreover, one can prove the following theorems, quoted as Theorems 7.1 and 7.2 in [1]:

Theorem *The quotient space $X^*/S^\perp$ is isometrically isomorphic to S^*. Furthermore, for each fixed $\phi \in X^*$,*

$$\sup_{x \in S, \, \|x\| \leq 1} |\phi(x)| = \min_{\psi \in S^\perp} \|\psi + \phi\|, \tag{3.36}$$

where "min" indicates that the infimum is attained.

Theorem *The space $(X/S)^*$ is isometrically isomorphic to $S^\perp$. Furthermore, for each fixed $y \in X$,*

$$\max_{\psi \in S^\perp, \, \|\psi\| \leq 1} |\psi(y)| = \inf_{x \in S} \|x + y\|, \tag{3.37}$$

where "max" indicates that the supremum is attained.

The above minimum norm problems represent the first examples of *duality theorems*, which relate an extremal problem in a space to another problem in the dual space. As remarked in [6], often the transition from one problem to its dual produces a significant simplification or enhances physical and mathematical insight. Sometimes, infinite-dimensional problems can be converted to equivalent finite-dimensional problems by considering the dual problem.

In the next section we shall write down the main duality theorem for H^p spaces, and in Sect. 3.7 we shall give a generalization to minimum norm problems for convex sets.

3.6 Extremal Problems in H^p Spaces

First we use the above results in order to obtain a representation of the dual space of H^p. We first recall a theorem of Riesz [1], which states that every bounded linear functional on L^p $(1 \leq p < \infty)$ has a unique representation

$$\phi(f) = \frac{1}{2\pi} \int_0^{2\pi} f(e^{i\theta}) g(e^{i\theta}) d\theta, \qquad g \in L^q, \tag{3.38}$$

where $1/p + 1/q = 1$. Moreover, $\|\phi\| = \|g\|_q$ and $(L^p)^*$ is isometrically isomorphic to L^q.

As discussed in Sect. 3.4, for $1 \leq p \leq \infty$ the set of boundary functions of H^p is the subspace of L^p for which

$$\int_0^{2\pi} e^{in\theta} f(e^{i\theta}) \, d\theta = 0, \qquad n = 1, 2, \ldots \tag{3.39}$$

Since H^p is a subspace of L^p, theorem (3.36) tells us that the space $(H^p)^*$ is isometrically isomorphic to the quotient $(L^p)^*/(H^p)^{\perp}$. So, we have to find a description of the annihilator $(H^p)^{\perp}$.

From (3.38) and (3.39) it is easy to see that

$$\int_0^{2\pi} f(e^{i\theta})g(e^{i\theta})\,d\theta = 0, \tag{3.40}$$

for every function $f \in H^p$, if $g \in H^q$ with $1/p + 1/q = 1$ and $g(0) = 0$.

Thus, the relation (3.40) identifies the annihilator $(H^p)^{\perp}$ as H^q (modulo the condition at the origin, which plays no role in the present arguments). From the relation (3.36) it follows that $(H^p)^*$ is isometrically isomorphic to L^q/H^q, where $1/p + 1/q = 1$. This result can be formulated more specifically as the following theorem, quoted as Theorem 7.3 in [1]:

Theorem *For $1 \le p < \infty$, the dual space $(H^p)^*$ is isometrically isomorphic to L^p/H^q, where $1/p + 1/q = 1$. Furthermore, if $1 < p < \infty$, each $\phi \in (H^p)^*$ is representable in the form*

$$\phi(f) = \frac{1}{2\pi} \int_0^{2\pi} f(e^{i\theta})(g(e^{i\theta}))^*\,d\theta, \tag{3.41}$$

by a unique function $g \in H^q$ (recall that g^* is the complex conjugate of g), while each $\phi \in (H^1)^*$ can be represented as (3.41) by some $g \in L^{\infty}$.

For the duality result given below, it is useful to write the most general bounded linear functional on H^p ($0 \le p < \infty$) in the form

$$\phi(f) = \frac{1}{2\pi i} \int_{|z|=1} f(z)\,k(z)\,dz, \tag{3.42}$$

where $k(e^{i\theta}) \in L^q$, $1/p + 1/q = 1$.

A function $h \in L^q$ is said to be *equivalent* to the given kernel k (which is written as $h \sim k$) if h and k belong to the same coset of L^q/H^q, that is if $h - k \in H^q$. Thus h and k define the same functional on H^p if and only if $h \sim k$.

Then for a fixed $k \in L^q$, a typical extremal problem is to find the norm in the dual space:

$$\|\phi\| = \sup_{f \in H^p,\, \|f\|_p \le 1} |\phi(f)|. \tag{3.43}$$

Besides the value of the supremum, it is of interest to know if the supremum is attained and if it is unique.

It is convenient to use the term *extremal function* to indicate a solution to the extremal problem (3.43). Since $|\phi(e^{i\alpha}f)| = |\phi(f)|$ for every real α, some normalization must be imposed before the extremal function can be unique. It is convenient to single out the extremal functions for which $\phi(f) = \|\phi\|$, which will be called *normalized* extremal functions.

The main results on existence and uniqueness are contained in the following theorem, quoted as Theorem 8.1 in [1]:

Theorem *For each p $(0 \leq p \leq \infty)$ and for each function $k(e^{i\theta}) \in L^q$ $(1/p + 1/q = 1)$ with $k \notin H^q$, the relation*

$$\|\phi\| = \sup_{f \in H^p, \|f\|_p \leq 1} |\phi(f)| = \min_{g \in H^q} \|k - g\|_q \qquad (3.44)$$

holds, where $\phi(f)$ is defined by (3.42). The notation indicates that the minimum on the r.h.s. is attained. If $p > 1$, there is a unique extremal function f for which $\phi(f) > 0$. If $p = 1$ and $k(e^{i\theta})$ is continuous, at least one extremal function exists. If $p > 1$ $(q < \infty)$, the dual extremal problem has a unique solution. If $p = 1$ $(q = \infty)$, the dual extremal problem has at least one solution; it is unique if an extremal function exists.

The equality (3.44) is a *duality relation*. It connects the original extremal problem (3.43) with what is called the *dual extremal problem*: to find the function $g \in H^q$ which is closest to the given kernel $k \in L^q$ (or, equivalently, to find the function $h \sim k$ of minimum norm). The proof of existence is based on the theorems (3.36) and (3.37) given in the previous section, while the uniqueness is obtained from a more detailed analysis [1].

An interesting relation exists among the extremal function $F \in H^p$ of the left hand side of (3.44), with $\|F\|_p = 1$ and $\phi(F) > 0$, and the extremal kernel K of the right hand side problem. We first note that

$$|\phi(F)| = \left| \frac{1}{2\pi} \int_{|z|=1} F(z) K(z) \, dz \right| \leq \frac{1}{2\pi} \int_0^{2\pi} |e^{i\theta} F(e^{i\theta}) K(e^{i\theta})| \, d\theta. \qquad (3.45)$$

But according to Hölder's inequality [2]

$$\frac{1}{2\pi} \int_0^{2\pi} |e^{i\theta} F(e^{i\theta}) K(e^{i\theta})| d\theta \leq \|F\|_p \|K\|_q = \|K\|_q, \qquad (3.46)$$

where in the last step we used $\|F\|_p = 1$.

The duality relation $\phi(F) = \|K\|_q$ requires to have equality in both relations (3.45) and (3.46). The equality in (3.45) implies

$$e^{i\theta} F(e^{i\theta}) K(e^{i\theta}) \geq 0, \quad \text{a. e.}, \qquad (3.47)$$

while the equality in Hölder's inequality (3.46) holds if and only if a constant $c \neq 0$ exists such that $|F(e^{i\theta})|^p = c|K(e^{i\theta})|^q$ [2]. This implies in particular that

$$|K(e^{i\theta})| = \|K\|_\infty \quad \text{a. e.} \quad \text{if} \quad p = 1. \qquad (3.48)$$

The duality theorem is very useful for the *minimal interpolation* problems, which require to find a function of minimal norm which takes prescribed values at some points z_i, $i = 1, \ldots n$, inside $|z| < 1$. This problem belongs to the so-called Nevanlinna–Pick class of interpolation problems. A related problem, known as Carathéodory-Fejér-Schur interpolation, is to find a function $f \in H^p$ of minimum norm whose first n derivatives $f^{(k)}(\beta)$, $k = 1, 2, \ldots, n$ at a point β in $|z| < 1$ have prescribed values.

The Nevanlinna–Pick problem can be formulated in the general case of an infinite number of points, when restrictions on the sequences $\{z_n\}$ and $\{w_n\}$ must be found in order to prove the existence of a H^p function such that $f(z_n) = w_n$. A detailed discussion of the general interpolation theory on H^p spaces is given in [1].

3.7 Geometrical View of Duality in Extremal Problems

For completeness, in this section we give an intuitive geometrical interpretation of the duality relations between an extremal problem in a vector space and another problem in the dual space. We also extend the duality theorem to minimum distance problems to convex sets.

We introduce first some useful definitions [6]. A *linear variety* V is defined as the translation of a subspace, i.e. it can be written as $V = x_0 + M$, where $x_0 \in X$ and M is a subspace (we used above the notion of *coset of X modulo M* for the same quantity). In this representation, the subspace M is unique, but any vector in V can serve as x_0.

A maximal proper variety in the linear space X, that is, a linear variety H such that $H \neq X$ and if V is any linear variety containing H, then either $V = H$ or $V = X$, is called a *hyperplane*. As illustrated in Fig. 3.1, while a subspace M is a generalization of our intuitive notion of a plane or a line containing the origin, a hyperplane H is a translation of a subspace and may not contain the origin. If the space X is normed, a hyperplane H must be either closed or dense in X: indeed, since H is a maximal linear variety, its closure $\bar{H}$ must coincide either with H or with X. Of interest in the theory of extremal problems are the closed hyperplanes in a normed vector space X.

This definition of hyperplanes is made without explicit references to linear functionals and emphasizes their geometrical interpretation. However, hyperplanes are related to linear functionals, as demonstrated by the following proposition [6]:

Fig. 3.1 A hyperplane H is
the translation of a subspace
M

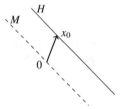

Fig. 3.2 Duality for the
minimum distance to a
convex set

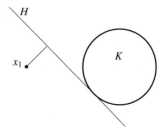

Proposition *Let H be a hyperplane in a normed linear space X. Then there is a linear functional f on X and a constant c, such that $H = \{x : f(x) = c\}$. Conversely, if f is a linear functional on X, the set $\{x : f(x) = c\}$ is a hyperplane in X. In particular, if H does not contain the origin, there is a unique linear functional f on X such that $H = \{x : f(x) = 1\}$. Furthermore, the hyperplane H is closed if and only if the functional f is continuous.*

The correspondence between a closed hyperplanes and an element of the dual space X^* suggests that all the concepts in which X^* plays a fundamental role can be visualized in terms of closed hyperplanes. This approach is followed in the presentation of the optimization problems in Ref. [6]. This allows in particular the generalization of the minimum norm problems from subspaces to convex sets.

Recall that a set K in a linear space X is *convex* if, given x_1, $x_2 \in K$, all points of the form $\alpha x_1 + (1 - \alpha)x_2$ with $0 \leq \alpha \leq 1$ are in K. To a given convex set, one can associate the important notion of supporting hyperplane [6].

We first note that for a nonzero linear functional on a linear normed space X, we can define, along with the hyperplane $H = \{x : f(x) = c\}$, also the closed *half-spaces* defined by the inequalities $\{x : f(x) \leq c\}$ or $\{x : f(x) \geq c\}$. Then a closed hyperplane H in a normed space X is said to be a *supporting hyperplane* for the convex set K if K is contained in one of the closed half-spaces determined by H and H contains a point of K. An obvious intuitive result is expressed by the following theorem [6]:

Theorem *If x is not an interior point of a convex set K which contains interior points, there is a closed hyperplane H containing x such that K lies on one side of H. Moreover, if K is closed, then it is equal to the intersection of all the closed half-spaces that contain it.*

We now consider the problem of finding the minimum distance from a point to a closed convex set K in a normed space X. The fundamental duality theorem is illustrated geometrically in Fig. 3.2: the minimum distance from a fixed point x_1 to a convex set K is equal to the maximum of the distances from the point to the hyperplanes H separating the point and the convex set K.

In order to translate this simple, intuitive, geometric relation into algebraic terms, we need the following definition: given a convex set K in a normed linear space X, the functional

$$h(x^*) = \sup_{x \in K} \langle x, x^* \rangle \qquad (3.49)$$

defined on X^* is called the *support functional of K*. Geometrically, it corresponds to a *supporting hyperplane H* such that K is contained in one of the closed half-spaces determined by H and H contains a point of the closure $\bar{K}$. We used here the symmetric notation of the dual space given in (3.33).

We can now formulate the minimum norm duality as the following theorem [6]:

Theorem *Let x_1 be a point in a real normed vector space X and let $d > 0$ denote its distance from the convex set K having the support functional h. Then*

$$d = \inf_{x \in K} \|x - x_1\| = \max_{|x^*\| \leq 1} [\langle x_1, x^* \rangle - h(x^*)], \qquad (3.50)$$

where the maximum on the right is achieved by some $x_0^* \in X^*$. If the infimum on the left is achieved by some $x_0 \in K$, then $-x_0^*$ is alligned with $x_0 - x_1$, that is the relation $\langle x_0 - x_1, -x_0^* \rangle = 0$ holds.

It is instructive to consider the case when $K = M$ is a subspace of the normed space X. Then, the supporting functional $h(x^*)$ is finite only for $x^* \in M^\perp$, in which case is zero. Therefore, we obtain from (3.50)

$$\inf_{m \in M} \|x_1 - m\| = \max_{|x^*\| \leq 1, \ x^* \in M^\perp} |\langle x_1, x^* \rangle|, \qquad (3.51)$$

which is equivalent to the earlier result (3.37), written in the notation defined in (3.33).

3.8 Generalized Lagrange Multipliers

In the previous sections we discussed a special case of extremal problems, in which one had to find the minimum distance from a fixed point to a subspace or a convex set in a linear space X. This was equivalent to a minimum norm over a certain class of vectors in X. However, in many optimization problems one is confronted with the problem of finding the extremal values of more general real functionals. Moreover, the subspace or the convex set involved in the optimization is often defined implicitly by means of a set of constraints. In these cases, the technique of Lagrange multipliers proves to be the best approach. In the most general case, a Lagrange multiplier is an element of the dual of a linear normed space [6].

First, we remark that by introducing a cone defining the positive vectors in a given space, it is possible to consider inequality problems in abstract vector spaces. Let P be a convex cone in a vector space X (recall that a set C in a linear space X is said to be a *cone with vertex at the origin* if $x \in C$ implies that $\alpha x \in C$ for all $\alpha \geq 0$). For $x, y \in X$ we write $x \geq y$ (with respect to P) if $x - y \in P$. The cone P defining this relation is called the *positive cone* in X. In the case of a normed space, we say that x is a positive vector and write $x > \theta$ if x is an interior point of the positive cone P.

There are several theorems which allow to find the solution of a constrained optimization problem by relating it to an unconstrained optimization of a modified functional, denoted as the Lagrange functional. For instance, consider the problem of finding

$$\inf_{x \in K, G(x) \le \theta} f(x) \tag{3.52}$$

where K is a convex subset of a linear space X, f is a real-valued convex functional on K and G is a convex mapping from K to a normed space Z having positive cone P (recall that a real functional f is *convex* if it satisfies the relation $f(\alpha x_1 + (1 - \alpha)x_2) \le \alpha f(x_1) + (1 - \alpha) f(x_2)$ for any $x_1, x_2 \in X$ and $0 \le \alpha \le 1$; a similar definition holds in the more general case of a mapping G to a normed space Z, if a positive cone P was defined in Z).

Given the problem (3.52), we attach to it the Lagrangian

$$L(x, z^*) = f(x) + \langle G(x), z^* \rangle, \tag{3.53}$$

where the Lagrange multiplier is some specific element $z^* \in Z^*$. From the experience with simple Lagrange multipliers, we expect that the solution of the original problem (3.52) will be related to the unconstrained minimum of $L(x, z^*)$ at a fixed z^*.

In fact, there are several theorems which express this result, belonging to both global and local optimization theory. For illustration, we quote below an important theorem from global optimization [6], which is the most modern approach of Lagrange theory. Starting from the problem (3.52), we define first the *dual functional* $\varphi(z^*)$ as

$$\varphi(z^*) = \inf_{x \in K} [f(x) + \langle G(x), z^* \rangle]. \tag{3.54}$$

In general, φ is not finite throughout the positive cone in Z^* but the region where it is finite is convex [6]. Then the following result, referred to in [6] as *Lagrange duality theorem*, relates the problems (3.52) and (3.54):

Theorem *Let f be a real-valued convex functional defined on a convex subset K of a vector space X and let $G(x)$ be a convex mapping of X into a normed space Z. Suppose there exists an x_1 such that $G(x_1) < \theta$ and $\mu_0 = \inf\{f(x) : G(x) \le \theta, x \in K\}$ is finite. Then*

$$\inf_{x \in K, \ G(x) \le \theta} f(x) = \max_{z^* \ge \theta} \varphi(z^*), \tag{3.55}$$

where the maximum on the right is achieved by some $z_0^* \ge \theta$. If the infimum on the left is achieved by some $x_0 \in K$, then the following allignment *condition holds:*

$$\langle G(x_0), z_0^* \rangle = 0 \tag{3.56}$$

and x_0 minimizes the functional $f(x) + \langle G(x), z_0^* \rangle$ for $x \in K$.

In practical applications one resorts usually to the so-called local theory of Lagrange multipliers, which is valid in the case when both the functional f and

the mapping G are smooth enough and can be differentiated. Then, one can implement the content of the global theory in the form of differential equations around
the stationary points of the Lagrangian. The local form of Lagrange optimization,
presented in detail in [6], is useful especially when the constraints are formulated as
an abstract equality $G(x) = \theta$ in the normed space Z.

In the next chapter we will discuss several specific optimization problems inspired
from physical situations, where we will make use of the mathematical notions presented in this chapter.

References

1. P.L. Duren, *Theory of H^p Spaces* (Academic Press, New York, 1970)
2. W. Rudin, *Real and Complex Analysis* (McGraw-Hill, New York, 1966)
3. B. Sz.-Nagy, C. Foias, *Harmonic Analysis of Operators on Hilbert Space* (Akadémiai
   Kiadó/North-Holland, Budapest/Amsterdam/London, 1970)
4. V.P. Potapov, Trudy Moskov. Mat. Obsc. **4**, 125 (1955). (in Russian)
5. R.E. Edwards, *Functional Analysis: Theory and Applications* (Holt, Rinehart and Winston, New
   York, 1965)
6. D.G. Luenberger, *Optimization by Vector Space Methods* (Wiley, New York, 1968)

Chapter 4
Optimization Problems with Physical Conditions

In this chapter, the general mathematical theory presented in the previous chapter is applied to problems encountered in hadron physics. We start by bringing a generic physical situation to the canonical form, by performimg the conformal mapping of the analyticity domain onto the interior of the unit disk. Then we explain the procedure of constructing an outer function from its modulus on the boundary for a typical case encountered in many physical situations. For completeness, we also briefly discuss the construction in the case of matrix-valued analytic functions. Several optimization problems of physical interest are then treated, using the technique of Lagrange multipliers and the duality theorem in L^2 and L^∞ norms. We discuss also the method of conformal mapping for extending the convergence domain and improving the convergence rate of power series. The chapter ends with a brief review of applications, presented in a historical perspective.

4.1 Conformal Mapping to the Unit Disk

A *conformal mapping* transforms two oriented intersecting curves from one complex plane to another complex plane, such that it preserves the angle between them in magnitude and in orientation. This means that the angle between two curves in the original plane will be identical to the angle between the corresponding curves in the second plane, although the transformed curves may not be similar to the original ones. A known result in complex analysis [1] is that a holomorphic function $F(z)$ is conformal at every point z where $F'(z) \neq 0$.

A mapping is called *biholomorphic* if it is a holomorphic bijection. It is known that, if F is a biholomorphic mapping, then F' is nowhere vanishing. Therefore, the term "conformal" in this context is synonymous with "biholomorphic", as the conditions are equivalent in the complex plane. A known result in complex analysis is *Riemann mapping theorem*, which states that every simply connected open set is biholomorphically equivalent to the unit disk [1].

For the problems studied in the most part of this chapter, the conformal mappings are suitable since they bring the original physical problem to a canonical form,

I. Caprini, *Functional Analysis and Optimization Methods in Hadron Physics*,
SpringerBriefs in Physics, https://doi.org/10.1007/978-3-030-18948-8_4

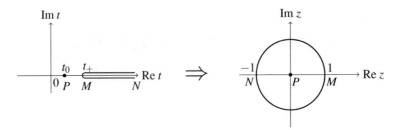

Fig. 4.1 Conformal mapping of the t-plane cut for $t \geq t_+$ onto the unit disk $|z| < 1$

allowing the application of standard techniques of functional optimization [2, 3], reviewed in the previous chapter.[1] By Riemann's theorem, the typical analyticity domains of the scattering amplitudes or form factors, reviewed in Chap. 1, can be mapped onto a unit disk. We shall consider below two particular cases encountered in applications (for a comprehensive list of conformal mappings see Ref. [4]).

We assume first that the original analyticity domain is the complex t plane cut along the real axis for $t \geq t_+$ (Fig. 4.1, left). As we saw in Sect. 1.3, this is the typical analytic structure of the hadronic form factors.

We define the conformal mapping

$$z \equiv \tilde{z}(t, t_0) = \frac{\sqrt{t_+ - t_0} - \sqrt{t_+ - t}}{\sqrt{t_+ - t_0} + \sqrt{t_+ - t}}, \tag{4.1}$$

where t_0 is the coordinate of an arbitrary point P situated inside the analyticity domain. For definiteness, we shall take P on the real axis below the branch point, i.e. we shall assume that $t_0 < t_+$.

By the change of variable (4.1), the t plane cut for $t \geq t_+$ is mapped onto the unit disk $|z| < 1$ in the z plane (Fig. 4.1, right), such that $\tilde{z}(t_0, t_0) = 0$, $\tilde{z}(t_+, t_0) = 1$ and $\tilde{z}(\infty, t_0) = -1$. One can easily check that the upper (lower) edge of cut $t \geq t_+$ becomes the unit semicircle $\zeta = e^{i\theta}$ with $\theta > 0$ ($\theta < 0$). From (4.1) one obtains also the inverse transformation

$$\tilde{t}(z, t_0) = 4t_+ \frac{z}{(1+z)^2} + t_0 \frac{(1-z)^2}{(1+z)^2}, \tag{4.2}$$

which maps the unit disk $|z| < 1$ onto the t plane cut along $t \geq t_+$.

In Sect. 1.2, we saw that in the case of scattering amplitudes the typical analyticity domain is the complex energy plane with two cuts along the real axis. For instance, Fig. 1.4 shows the cuts in the s plane for an amplitude $T(s, t)$ as a function of s at fixed t. A similar structure with a right and a left cut is exhibited also by the partial waves $t_l(s)$, and we shall encounter again this structure for another function of interest in Sect. 6.3.

---

[1]The conformal mappings turn out to be useful also in a different context, which will be discussed later in Sect. 4.9.

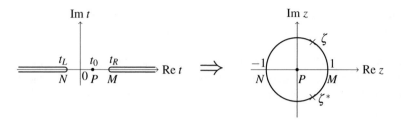

Fig. 4.2 Conformal mapping of the t plane cut for $t \geq t_R$ and $t \leq t_L$ onto the unit disk $|z| < 1$

In Fig. 4.2, left, we represent a typical situation in a generic complex plane t, with a cut for $t \geq t_R$ and a cut for $t \leq t_L$. In this case we shall define the conformal mapping

$$z \equiv \tilde{z}(t, t_0) = \frac{\sqrt{t - t_L} - a\sqrt{t_R - t}}{\sqrt{t - t_L} + a\sqrt{t_R - t}}, \qquad a = \frac{\sqrt{t_0 - t_L}}{\sqrt{t_R - t_0}}, \qquad (4.3)$$

where t_0 is the coordinate of an arbitrary point P situated inside the analyticity domain, which we shall take again on the real axis between the branch points, $t_L < t_0 < t_R$.

By the conformal mapping (4.3), the t plane cut for $t \geq t_R$ and $t \leq t_L$ is mapped onto the unit disk $|z| < 1$ in the z plane (Fig. 4.2, right), such that $\tilde{z}(t_0, t_0) = 0$, $\tilde{z}(t_R, t_0) = 1$ and $\tilde{z}(t_L, t_0) = -1$. The upper (lower) edge of cuts $t \geq t_R$ and $t \leq t_L$ becomes the unit semicircle $\zeta = e^{i\theta}$ with $\theta > 0$ ($\theta < 0$), and the points $t \to \infty$ correspond to two points on the unit circle, ζ and ζ^*, where

$$\zeta = \frac{1 + ai}{1 - ai}. \qquad (4.4)$$

From (4.3) one obtains the inverse transformation

$$\tilde{t}(z, t_0) = \frac{a^2 t_R (1 + z)^2 + t_L (1 - z)^2}{a^2 (1 + z)^2 + (1 - z)^2}. \qquad (4.5)$$

By the conformal mappings of the form (4.1) or (4.3), the problems of physical interest can be brought to a canonical form. Therefore, in this chapter we shall treat only optimization problems formulated in the unit disk $|z| < 1$.

4.2 Construction of Outer Functions

We defined in Sect. 3.2 an outer function as a function analytic and without zeros in $|z| < 1$, which is obtained from its modulus on the boundary through the integral representation (3.15). The outer functions appear in the product (3.18) expressing the factorization theorem for the functions of H^p class. They will play an important role in the optimization problems discussed in this chapter.

It turns out that in many cases of physical interest the outer functions can be constructed explicitly as products of several factors, obtained by a simple recipe. We illustrate this by considering a hadronic form factor $F(t)$ analytic in the domain shown in Fig. 4.1 left, and which satisfies a boundary integral condition of the form

$$\frac{1}{\pi} \int_{t_+}^{\infty} w(t) \, |F(t)|^2 \, dt \leq 1, \tag{4.6}$$

where $w(t) \geq 0$ is a known weight. We shall explain in Chap. 5 how such model-independent constraints on form factors can be derived. After the conformal mapping (4.1), the inequality (4.6) becomes

$$\frac{1}{2\pi} \int_{0}^{2\pi} w(\tilde{t}(\zeta, t_0)) \left| \frac{d\tilde{t}(\zeta, t_0)}{d\zeta} \right| |F(\tilde{t}(\zeta, t_0))|^2 d\theta \leq 1, \quad \zeta = e^{i\theta}, \tag{4.7}$$

where $\tilde{t}$ is the inverse transformation defined in (4.2).

In order to find the optimal solution of many extremal problems on the class of analytic functions $F(t)$ satisfying the condition (4.7), it is necessary to write this condition in the equivalent form

$$\frac{1}{2\pi} \int_{0}^{2\pi} |\phi(e^{i\theta})|^2 \, |F(\tilde{t}(e^{i\theta}, t_0))|^2 \, d\theta, \tag{4.8}$$

where $\phi(z)$ is an outer function, i.e. a function without zeros in $|z| < 1$, whose modulus on the boundary, $|\zeta| = 1$, is given by

$$|\phi(\zeta)|^2 = w(\tilde{t}(\zeta, t_0)) \left| \frac{d\tilde{t}(\zeta, t_0)}{d\zeta} \right|. \tag{4.9}$$

One might ask why $\phi(z)$ should be an outer function. We shall give here an heuristic explanation (more formal arguments can be found in the mathematical literature, for instance [2]). Actually, it is possible to insert in (4.8) an inner function without changing the modulus, since by definition the modulus of an inner function is equal to 1 on the boundary. Disregarding singular functions,[2] we can insert a Blaschke product of the form (3.14). However, since the Blaschke factors vanish at some points inside the analyticity domain, they will lead to weaker constraints or even to no constraints on the form factor at those points or near them. So, optimal solutions of the extremal problems can be found only by replacing the boundary values by the modulus of an outer function.

---

[2]A singular inner function of the form (3.16) contains a phase which infinitely oscillates, a feature assumed not to occur in the case of physical amplitudes.

As we mentioned, an outer function is obtained inside the unit disk from its modulus on the boundary by the integral representation (3.15). In many physical cases, the weight $w(t)$ is a product of simple factors of the form

$$w(t) = \left(1 - \frac{t_+}{t}\right)^k (t - t_-)^l t^m (t + Q^2)^n, \qquad (4.10)$$

where t_- and Q^2 are positive momenta squared and k, l, m, n are real (positive or negative) numbers, which may be nonintegers. As we shall show below, for weights of the form (4.10), the integration (3.15) can be avoided, and the outer function $\phi(z)$ satisfying the condition (4.9) can be obtained by a straightforward procedure.

It can be shown that the solutions of the extremal problems based on the condition (4.6) are independent on the parameter t_0 in (4.1), i.e. on the choice of the particular conformal mapping onto the unit disk. On the other hand, in form-factor parametrizations based on truncated expansions in powers of the variable z, the different expansions are no longer equivalent and an optimal choice will be adopted. Therefore, it is necessary to construct the outer functions for arbitrary values of t_0. There are several ways to do this. We choose to perform first the calculations for $t_0 = 0$ and make at the end the transformations which will give the general result.

We note first that each factor in (4.10) and the Jacobian $|d\tilde{t}/dz|$ can be continued from the boundary function to a function analytic in $|z| < 1$. Thus, for $t_0 = 0$, we obtain from (4.1)

$$\frac{d\tilde{t}(z, 0)}{dz} = 4t_+ \frac{1 - z}{(1 + z)^3}, \qquad (4.11)$$

and

$$1 - \frac{t_+}{\tilde{t}(z, 0)} = \left(\frac{1 - z}{1 + z}\right)^2. \qquad (4.12)$$

These functions are analytic and do not vanish for $|z| < 1$, therefore the corresponding outer functions are obtained by simply taking the aquare root of the factors in the r.h.s:

$$\left|\frac{d\tilde{t}(z, 0)}{dz}\right| \Rightarrow \phi_1(z) = \sqrt{4t_+} \frac{(1 - z)^{1/2}}{(1 + z)^{3/2}},$$

$$1 - \frac{t_+}{t} \Rightarrow \phi_2(z) = \frac{1 - z}{1 + z}. \qquad (4.13)$$

Incidentally, the function $\phi_1(z)$ belongs to H^p with $p < 2/3$ and $\phi_2(z)$ belongs to H^p with $p < 1$.

The remaining functions in the product (4.10) vanish inside the analyticity domain, i.e. the t plane cut for $t \geq t_+$. To remove the zeros without changing the modulus on the boundary $|z| = 1$, it is convenient to make use of a Blaschke function (3.14) of the form

$$B(z, \xi) = \frac{z - \xi}{1 - \xi^* z},$$ (4.14)

which vanishes at a suitable point ξ, $B(\xi, \xi) = 0$.

We treat first the factor $t \equiv \tilde{t}(z, 0)$. From (4.2), it follows that $\tilde{t}(z, 0)$ is analytic in $|z| < 1$ and vanishes at $z = 0$. This zero can be eliminated simply by removing the factor $z \equiv B(z; 0)$ from (4.2). Clearly, this does not change the modulus, since the removed factor is unimodular on $|z| = 1$. Thus, the nonvanishing function corresponding to t is

$$4t_+ \frac{1}{(1 + z)^2},$$ (4.15)

which means that the outer function associated to the factor t will be

$$t \quad \Rightarrow \quad \phi_3(z) = \sqrt{4t_+} \frac{1}{1 + z}.$$ (4.16)

We consider now the factor $t + Q^2$. By using (4.2) with $t_0 = 0$, it writes as

$$t + Q^2 = 4t_+ \left[\frac{z}{(1 + z)^2} - \frac{\tilde{z}(-Q^2, 0)}{(1 + \tilde{z}(-Q^2, 0))^2} \right],$$ (4.17)

or, equivalently as the expression

$$t + Q^2 = 4t_+ \frac{(z - \tilde{z}(-Q^2, 0))(1 - z\tilde{z}(-Q^2, 0)}{(1 + z)^2 (1 + \tilde{z}(-Q^2, 0))^2},$$ (4.18)

which exhibits a zero at $z = \tilde{z}(-Q^2, 0)$. The zero can be eliminated if we divide the above expression by the Blaschke factor (4.14) with $\xi = \tilde{z}(-Q^2, 0)$, which amounts to replacing $t + Q^2$ by

$$4t_+ \frac{(1 - z\tilde{z}(-Q^2, 0)^2}{(1 + z)^2 (1 + \tilde{z}(-Q^2, 0))^2}.$$ (4.19)

Using (4.1) with $t_0 = 0$, we can write

$$\tilde{z}(-Q^2, 0) = \frac{1 - \beta_Q}{1 + \beta_Q}, \quad \text{where} \quad \beta_Q = \sqrt{1 + \frac{Q^2}{t_+}}.$$ (4.20)

With this notation, we obtain from (4.19) the outer function corresponding to the factor $t + Q^2$:

$$t + Q^2 \quad \Rightarrow \quad \phi_4(z) = \sqrt{t_+} \left(\frac{1 - z}{1 + z} + \beta_Q \right).$$ (4.21)

From this we obtain the similar correspondence

$$t - t_- \quad \Rightarrow \quad \phi_5(z) = \sqrt{t_+} \left(\frac{1-z}{1+z} + \beta_- \right), \quad \beta_- = \sqrt{1 - \frac{t_-}{t_+}}. \tag{4.22}$$

The full outer function appearing in (4.8) can be finally written down by collecting the contributions given in (4.13), (4.16), (4.21) and (4.22). The exponents appearing in (4.10) can be added without problems, even if they are not integers, since the functions $\phi_i(z)$ have no zeros which might produce branch points inside the unit disk. So, we can write

$$\phi(z) = \phi_1(z)\phi_2(z)^k \phi_5(z)^l \phi_3(z)^m \phi_4(z)^n, \tag{4.23}$$

which gives

$$\phi(z) = 2^{m+1} t_+^{(l+m+n+1)/2} \frac{(1-z)^{k+1/2}}{(1+z)^{k+m+1/2}} \left(\frac{1-z}{1+z} + \beta_- \right)^l \left(\frac{1-z}{1+z} + \beta_Q \right)^n. \tag{4.24}$$

We now briefly discuss how to obtain the outer function in the general case when the parameter t_0 in (4.1) is not zero. We denote as above $z \equiv \tilde{z}(t, 0)$ and let z' be the variable $z' \equiv \tilde{z}(t, t_0)$. By definition $\tilde{z}(t_0, t_0) = 0$ and from (4.1) we have

$$\tilde{z}(t_0, 0) = \xi, \quad \xi \equiv \frac{1 - \sqrt{1 - t_0/t_+}}{1 + \sqrt{1 - t_0/t_+}} < 1. \tag{4.25}$$

Therefore, the variables z and z' are related by the Blaschke transformations

$$z' = \frac{z - \xi}{1 - \xi z}, \quad z = \frac{z' + \xi}{1 + \xi z'}, \tag{4.26}$$

which satisfy

$$\frac{dz}{dz'} = \frac{1 - \xi^2}{(1 + \xi z')^2}. \tag{4.27}$$

In order to obtain the outer function for the variable z', we write $d\theta = |d\zeta/d\zeta'| \, d\theta'$ and use (4.27), which brings (4.8) to

$$\frac{1}{2\pi} \int_0^{2\pi} |\phi'(\zeta')|^2 \, |F|^2 \, d\theta', \tag{4.28}$$

where

$$\phi'(z') = \phi(z) \frac{\sqrt{1 - \xi^2}}{(1 + \xi z')}, \quad z = \frac{z' + \xi}{1 + \xi z'}. \tag{4.29}$$

This relation gives the recipe to obtain the outer function for the general mapping (4.1), starting from the outer function in the particular case $t_0 = 0$.

4.3 Matrix-Valued Outer Functions

This section contains a short discussion of the more general case of matrix-valued functions, which appear when we consider simultaneously several physical amplitudes. An example of boundary condition involving a complex matrix-valued function was given in Sect. 2.2. The aim in this case is to write the physical boundary condition involving several amplitudes in a "diagonal" form, which is suitable for finding the optimal solution of the extremal problems for these amplitudes.

By a conformal mapping of the analyticity domain onto the unit disk $|z| < 1$, we obtain in these cases a boundary condition written generically in matrix form as

$$T^*(\zeta)P(\zeta)T(\zeta) = 1, \qquad |\zeta| = 1, \tag{4.30}$$

where $P(\zeta)$ is an $n \times n$ positive-definite matrix and $T(z) \equiv \{T_i(z)\}_1^n$ is a vector-valued analytic function in $|z| < 1$. According to the factorization theorem (3.19) given in Sect. 3.3, the matrix $P(\zeta)$ can be always expressed as a product

$$P(\zeta) = N^\dagger(\zeta)N(\zeta), \tag{4.31}$$

where $N(z)$ is an analytic matrix invertible in $|z| < 1$. This allows to define the new amplitudes

$$A_i(z) = \sum_{j=1}^{n} N_{ij}(z)T_j(z), \qquad i = 1, \ldots, n, \tag{4.32}$$

which are by construction analytic in $|z| < 1$ and equivalent to the original amplitudes T_i. Furthermore, the condition (4.30) takes the diagonal form

$$A^*(\zeta)A(\zeta) \equiv \sum_{i=1}^{n} A_i^*(\zeta)A_i(\zeta) = 1, \tag{4.33}$$

which is suitable in extremal problems.

As we mentioned below Eq. (3.19), there are no general recipes to obtain the matrix $N(\zeta)$ from a given $P(\zeta)$. However, in some physical cases $P(\zeta)$ is often naturally factorized as

$$P(\zeta) = Q^\dagger(\zeta)Q(\zeta), \tag{4.34}$$

where however the matrix $Q(\zeta)$ contains zeros or singularities (poles or branch points) when continued to $|z| < 1$. In the particular case when the matrix elements $Q_{ij}(\zeta)$ can be written as

$$Q_{ij}(\zeta) = \rho_i(\zeta)\tilde{Q}_{ij}(\zeta), \qquad i, j = 1, \ldots, n, \tag{4.35}$$

where the nonanalytic factors are contained in some known functions $\rho_i(\zeta) \geq 0$ and the matrix $\tilde{Q}$ is analytic, we can convert the factors ρ_i into outer functions G_i defined by the relations

$$|G_i(\zeta)| = \rho_i(\zeta), \qquad |\zeta| = 1, \qquad i, \dots, n. \tag{4.36}$$

The functions G_i can be constructed from the general integral relation (3.15), or using the technique described in Sect. 4.2 when ρ_i is a product of simple factors like those in (4.10). Then the matrix R defined by

$$R_{ij}(\zeta) = G_i(\zeta)\tilde{Q}_{ij}(\zeta), \qquad i, j = 1, \dots, n, \tag{4.37}$$

is analytic in $|z| < 1$ and satisfies the relation

$$R^\dagger(\zeta)R(\zeta) = Q^\dagger(\zeta)Q(\zeta), \tag{4.38}$$

so it can be used in the factorization

$$P(\zeta) = R^\dagger(\zeta)R(\zeta). \tag{4.39}$$

However, the problem is not yet fully solved, since the analytic matrix $R(z)$ may be not invertible, i.e. the condition det $R(z) \neq 0$ may be violated. Fortunately, the zeros of det $R(z)$ can be eliminated using the Blaschke-Potapov factorization discussed in Sect. 3.3. We sketch below the procedure assuming for simplicity that $n = 3$.

Let z_0 with $|z_0| < 1$ be a simple zero of det $R(z)$. The numerical matrix $R(z_0)$ can be reprezented according to known rules as

$$R(z_0) = UDV, \tag{4.40}$$

where U and V are both unitary matrices while D is diagonal. Since det $R(z_0) = 0$, one of the diagonal elements of D should be zero. Then D should be, say, of the form

$$D = \begin{bmatrix} 0 & 0 & 0 \\ 0 & d_2 & 0 \\ 0 & 0 & d_3 \end{bmatrix}. \tag{4.41}$$

Define the diagonal matrix

$$b(z) = \begin{bmatrix} \frac{z-z_0}{1-z_0^* z} & 0 & 0 \\ 0 & d_2 & 0 \\ 0 & 0 & d_3 \end{bmatrix} \tag{4.42}$$

which is unitary on the boundary of the unit disk

$$b^\dagger(\zeta)b(\zeta) = b(\zeta)b^\dagger(\zeta) = 1, \qquad |\zeta| = 1, \tag{4.43}$$

and is analytic inside the disk. Introduce then the Blaschke-Potapov matrix

$$B(z) = U b(z) V, \tag{4.44}$$

which is also unitary on the boundary

$$B^\dagger(\zeta) B(\zeta) = B(\zeta) B^\dagger(\zeta) = 1, \qquad |\zeta| = 1, \tag{4.45}$$

is analytic in $|z| < 1$ and is invertible everywhere, except the point z_0. Then, as shown in [5], the matrix

$$N(z) = B^{-1}(z) R(z) \tag{4.46}$$

is analytic in $|z| < 1$ and $\det N(z_0) \neq 0$. Since on the boundary $N(\zeta)$ differs from $R(\zeta)$ by an unitary matrix, it satisfies

$$N^\dagger(\zeta) N(\zeta) = R(\zeta) R^\dagger(\zeta), \qquad |\zeta| = 1. \tag{4.47}$$

Recalling the relation (4.39), it follows that by this construction we obtained the desired matrix $N(\zeta)$ appearing in the factorization (4.31).

4.4 Optimal Interpolation in L^2 Norm

We start the presentation of optimization problems useful in hadron physics by the so-called interpolation in L^2 norm. The problem can be formulated as follows: given a form factor which satisfies the boundary condition (4.6), find the optimal constraints satisfied by the values $F(t_n)$, $n = 1, \ldots, N$ at N points t_n inside the analyticity domain, and the K derivatives $F^{(k)}(t_0)$, $k = 1, \ldots, K$ at some point t_0.

In this section and in the following ones in this chapter we shall consider only analytic functions of real type, i.e. functions which satisfy the Schwarz reflection property $F(t^*) = F^*(t)$, since this is the case most encountered in applications.[3] Moreover, having in view the phenomenological applications, we shall restrict the discussion to constraints at points on the real axis, so that $t_n = t_n^*$ and $t_0 = t_0^*$.

We have seen in the previous sections that, after performing the conformal mapping (4.1) which maps the point t_0 to $z = 0$, and the definition of the outer function $\phi(z)$, the original condition (4.6) can be written in the equivalent form (4.8). Further, by introducing a new analytic function $g(z)$ defined as

$$g(z) = F(\tilde{t}(z, t_0)) \phi(z), \tag{4.48}$$

the boundary condition becomes

---

[3]As discussed in Sect. 1.3.2, this property is not satisfied by $\pi\omega$ form factor. This case will be considered in Sect. 5.5.

$$\frac{1}{2\pi} \int_0^{2\pi} d\theta |g(e^{i\theta})|^2 \leq 1, \tag{4.49}$$

showing that $g(z)$ is an the analytic function of class H^2. In the standard notation (3.5) of the L^p norm this condition is written as

$$\|g\|_2^2 \leq 1. \tag{4.50}$$

By expanding

$$g(z) = \sum_{n=0}^{\infty} g_k z^k, \tag{4.51}$$

where g_k are real, the condition (4.50) is written as

$$\sum_{k=0}^{\infty} g_k^2 \leq 1. \tag{4.52}$$

We consider the general case when the value of $g(z)$ and its first $K - 1$ derivatives at $z = 0$, as well as the values at N interior points $z_n \neq 0$ are known:

$$\left[\frac{1}{k!} \frac{d^k g(z)}{dz^k} \right]_{z=0} = g_k, \quad 0 \leq k \leq K - 1;$$
$$g(z_n) = \xi_n, \quad 1 \leq n \leq N, \tag{4.53}$$

where g_k and ξ_n are given numbers. They are related, by means of (4.48), to the derivatives $F^{(j)}(t_0)$, $j \leq k$ of $F(t)$ at a specific point $t = t_0$, and the values $F(\tilde{t}(z_n))$, respectively. Since we consider only real-analytic functions and assumed t_n to be real, it follows that z_n and ξ_n are also real. The problem which we consider is to characterize the allowed domain of the values $\{g_k, \xi_n\}$, consistent with the boundary condition (4.50).

As mentioned in the previous chapter, this mathematical problem is known as a general Carathéodory-Fejér-Schur-Pick-Nevanlinna interpolation [2]. In what follows, we shall show that in the case of the H^2 space an explicit solution of this problem can be found. To this end, we shall first show that the most general constraint satisfied by the input values appearing in (4.53) is given by the inequality:

$$\delta_2^2 \equiv \min_{g \in \mathcal{G}} \|g\|_2^2 \leq 1, \tag{4.54}$$

where the minimum is taken over the class $\mathcal{G}$ of analytic functions in H^2 which satisfy the conditions (4.53).

The proof of this assertion [6], starts from the remark that the values g_k and ξ_n consistent with the condition (4.50) form a closed and convex domain $\mathcal{D}$ in the real Euclidean space of dimension $K + N$. Let us take first a point inside this domain.

This means that there exists at least one function $g(z)$ of L^2 norm less than 1 which satisfies these specific constraints. Then it is sure that the minimum in the left side of (4.54) will be also less than 1. On the other hand, if a point with coordinates g_k and ξ_n is outside $\mathcal{D}$, the L^2 norms of all the functions $g(z)$ satisfying the corresponding constraints (4.53) will be larger than 1, and the same will be true by taking the minimum of these L^2 norms. By using the fact that the minimum is reached by some admissible function [2], it follows that the inequality (4.54) is a necessary and sufficient condition for the values g_k and ξ_n to belong to $\mathcal{D}$. Using the convexity of this domain, it follows also that its frontier is described by the values which achieve the equality in (4.54).

In the next subsections we shall solve the minimization problem (4.54) by two different methods.

4.4.1 Lagrange Multipliers

One may set up a Lagrangian for the minimization problem (4.54) with the constraints (4.53). By using the expansion (4.51), we can write the Lagrangian as

$$\mathcal{L} = \sum_{k=0}^{\infty} g_k^2 + 2 \sum_{n=1}^{N} \alpha_n \left(\bar{\xi}_n - \sum_{k=K}^{\infty} g_k z_n^k \right), \tag{4.55}$$

where α_n are Lagrange multipliers, and $\bar{\xi}_n$ are known numbers defined as

$$\bar{\xi}_n = \xi_n - \sum_{k=0}^{K-1} g_k z_n^k. \tag{4.56}$$

Solving the Lagrange equations obtained by varying $\mathcal{L}$ with respect to the free parameters g_k for $k \geq K$ gives

$$g_k = \sum_{n=1}^{N} \alpha_n z_n^k, \qquad k \geq K. \tag{4.57}$$

By imposing the constraints, i.e. the vanishing of the parantheses in (4.55) with g_k given by (4.57), we obtain

$$\sum_{m=1}^{N} \alpha_m \frac{z_m^K z_n^K}{1 - z_m z_n} = \bar{\xi}_n, \qquad n = 1, \ldots, N. \tag{4.58}$$

For the optimal g_k, the Lagrangian coincides with the minimum δ_2^2, which becomes

$$\delta_2^2 = \sum_{k=0}^{K-1} g_k^2 + \sum_{n=1}^{N} \sum_{m=1}^{N} \alpha_n \alpha_m \frac{z_n^K z_m^K}{1 - z_m z_n}. \tag{4.59}$$

By using again (4.58), the second term in the r.h.s. takes a simpler form, and we can express the inequality (4.54) as:

$$\sum_{n=1}^{N} \alpha_n \bar{\xi}_n \leq \bar{I}, \tag{4.60}$$

where we have defined

$$\bar{I} = 1 - \sum_{k=0}^{K-1} g_k^2. \tag{4.61}$$

Equations (4.58) and (4.60) can be viewed as a system of $N + 1$ linear relations for the N real parameters α_n. The consistency of this system implies the positivity of the determinant of the matrix of coefficients:

$$\det \begin{bmatrix} \bar{I} & \bar{\xi}_1 & \bar{\xi}_2 & \cdots & \bar{\xi}_N \\ \bar{\xi}_1 & \frac{z_1^{2K}}{1-z_1^2} & \frac{(z_1 z_2)^K}{1-z_1 z_2} & \cdots & \frac{(z_1 z_N)^K}{1-z_1 z_N} \\ \bar{\xi}_2 & \frac{(z_1 z_2)^K}{1-z_1 z_2} & \frac{(z_2)^{2K}}{1-z_2^2} & \cdots & \frac{(z_2 z_N)^K}{1-z_2 z_N} \\ \vdots & \vdots & \vdots & \vdots & \vdots \\ \bar{\xi}_N & \frac{(z_1 z_N)^K}{1-z_1 z_N} & \frac{(z_2 z_N)^K}{1-z_2 z_N} & \cdots & \frac{z_N^{2K}}{1-z_N^2} \end{bmatrix} \geq 0. \tag{4.62}$$

Moreover, all the principal minors of the above matrix should be nonnegative [7].

We note that in the above approach, the Lagrange multipliers have been used for implementing the conditions at the points z_n, while the derivatives at $z = 0$ have been implemented explicitly. Alternatively, as was done in Ref. [8], the solution can be obtained by introducing Lagrange multipliers also for the constraints on the given coefficients g_k, $k = 0, \ldots, K - 1$ in (4.55). This leads to the inequality:

$$\det \begin{bmatrix} 1 & g_0 & g_1 & \cdots & g_{K-1} & \bar{\xi}_1 & \bar{\xi}_2 & \cdots & \bar{\xi}_N \\ g_0 & 1 & 0 & 0 & 0 & 1 & 1 & \cdots & 1 \\ g_1 & 0 & 1 & 0 & 0 & z_1 & z_2 & \cdots & z_N \\ \vdots & \vdots & \vdots & \vdots & \vdots & \vdots & \vdots & \vdots & \vdots \\ g_{K-1} & 0 & 0 & \cdots & 1 & z_1^{K-1} & z_2^{K-1} & \cdots & z_N^{K-1} \\ \bar{\xi}_1 & 1 & z_1 & \cdots & z_1^{K-1} & \frac{1}{1-z_1^2} & \frac{1}{1-z_1 z_2} & \cdots & \frac{1}{1-z_1 z_N} \\ \bar{\xi}_2 & 1 & z_2 & \cdots & z_2^{K-1} & \frac{1}{1-z_1 z_2} & \frac{1}{1-z_2^2} & \cdots & \frac{1}{1-z_2 z_N} \\ \vdots & \vdots & \vdots & \vdots & \vdots & \vdots & \vdots & \vdots & \vdots \\ \bar{\xi}_N & 1 & z_N & \cdots & z_N^{K-1} & \frac{1}{1-z_1 z_N} & \frac{1}{1-z_2 z_N} & \cdots & \frac{1}{1-z_N^2} \end{bmatrix} \geq 0. \tag{4.63}$$

The equivalent conditions (4.62) and (4.63) are expressed in a straightforward way in terms of the values of the form factor $F(t)$ at $t_n = \tilde{t}(z_n)$ and the derivatives at $t = t_0$, using the conformal mapping (4.1) and the definition (4.48) of the function $g(z)$. It can be shown that they define convex domains in the space of the input parameters. One can prove also that, for a fixed weight $w(t)$ in (4.6), the bounds become stronger/weaker if the r.h.s. is replaced by a number smaller/larger than 1, respectively [9, 10].

The particular case when no derivatives at $z = 0$ appear is obtained by setting formally $K = 0$ in (4.62) and by omitting the lines and the columns starting with $g_0, \ldots, g_{K-1}$ in (4.63). Keeping for instance a single point z_1 in the determinants, we obtain from both these inequalities the constraint

$$|g(z_1)| \le \frac{1}{\sqrt{1 - z_1^2}}, \qquad |z_1| < 1. \tag{4.64}$$

Using (4.48), we obtain from this bound a constraint on the form factor itself

$$|F(t_1)| \le \frac{1}{|\phi(z_1)|} \frac{1}{\sqrt{1 - z_1^2}}, \tag{4.65}$$

where $t_1 = \tilde{t}(z_1, t_0)$ is a real point below the unitarity threshold t_+.

The opposite extreme case with no conditions at points $z_n \ne 0$ is obtained by removing from the determinants the lines and columns involving the quantities $\bar{\xi}_n$ or ξ_n. Then we obtain immediately from (4.62) the condition $\bar{I} \ge 0$, which implies

$$\sum_{k=0}^{K-1} g_k^2 \le 1, \tag{4.66}$$

where g_k appear in the expansion

$$F(t) = \frac{1}{\phi(z)} \sum_{k=0}^{K-1} g_k z^k, \qquad z = \tilde{z}(t, t_0). \tag{4.67}$$

The same condition is obtained from (4.63).

4.4.2 Analytic Approach

Instead of using Lagrange multipliers, we can alternatively follow a method where the constraints are explicitly taken into account in the expressions of the functions considered in the optimization. Since this amounts to an analytic implementation of the interpolation conditions, we refer to this approach as to an "analytic" interpolation.

We first define the Blaschke products

$$B(z) = \prod_{n=1}^{N} \frac{z - z_n}{1 - z_n^* z}, \qquad B_n(z) = \prod_{i \neq n}^{N} \frac{z - z_i}{1 - z_i^* z}. \qquad (4.68)$$

The crucial remark is that the conditions (4.53) can be fulfilled by expanding the functions $g \in \mathscr{G}$ in the most general way as[4]

$$g(z) = \sum_{k=0}^{K-1} g_k z^k + z^K \sum_{n=1}^{N} \frac{\bar{\xi}_n}{z_n^K} \frac{B_n(z)}{B_n(z_n)} - z^K B(z) f(z), \qquad (4.69)$$

where $\bar{\xi}_n$ are defined in (4.56). In this representation, the function $f(z)$ is analytic of class H^2 in $|z| < 1$ and is free of constraints. Expressed in terms of f, the minimum norm problem (4.54) becomes

$$\delta_2^2 = \min_{f \in H^2} \| f - h \|_2^2, \qquad (4.70)$$

where h is a function defined on the boundary of the unit disk $\zeta = \exp(i\theta)$ as

$$h(\zeta) = \frac{1}{\zeta^K B(\zeta)} \sum_{k=0}^{K-1} g_k \zeta^k + \frac{1}{B(\zeta)} \sum_{n=1}^{N} \frac{\bar{\xi}_n}{z_n^K} \frac{B_n(\zeta)}{B_n(z_n)}. \qquad (4.71)$$

In writing (4.70) we took into account the fact that the factor $z^N B(z)$ has modulus equal to 1 on the boundary, so we can devide by it in the expression of $|g(\zeta)|$.

The solution of (4.70) represents the minimum distance to the space H^2 of a given function $h \notin H^2$. We can calculate it by using the duality theorem (3.44) given in Sect. 3.6. For H^2, we can use actually a more straightforward method. We expand:

$$f(\zeta) = \sum_{l=0}^{\infty} f_l \zeta^l, \qquad (4.72)$$

and

$$h(\zeta) = \sum_{l=-\infty}^{\infty} h_l \zeta^l, \qquad (4.73)$$

where h_l are known real numbers defined as

---

[4]An expansion slightly different from (4.69), used in [6], leads to identical results.

$$h_l = \frac{1}{2\pi i} \int_{|\zeta|=1} \zeta^{-l} h(\zeta) \frac{d\zeta}{\zeta}, \qquad -\infty < l < \infty. \tag{4.74}$$

Using the expansions (4.72) and (4.73) and we write (4.70) as:

$$\delta_2^2 = \min_{\{f_l\}} \left[\sum_{l=0}^{\infty} (f_l - h_l)^2 + \sum_{l=-\infty}^{-1} h_l^2 \right], \tag{4.75}$$

where the minimization is taken upon the numbers f_l, which are free of constraints. Obviously, the minimum is reached for

$$f_l = h_l, \qquad l \geq 0, \tag{4.76}$$

which leads to

$$\delta_2^2 = \sum_{l=-\infty}^{-1} h_l^2. \tag{4.77}$$

The negative-frequency coefficients h_l for $l \leq -1$ are calculated by inserting into (4.74) the expression of h from (4.71) and by applying residue theorem. The poles are produced by the factors $B(z)$ and z^{K+l-k} for $K + l - k \geq 0$. By a straightforward calculation and using again (4.56), we write the contribution of $B(z)$ to h_l as

$$h_l \sim \sum_{n=1}^{N} \frac{1 - z_n^2}{B_n(z_n)} \frac{\xi_n}{z_n^{K+l+1}}, \tag{4.78}$$

while the factor ζ^{K+l-k} gives

$$h_l \sim \sum_{k=0}^{K-1} \theta(K + l - k) \frac{g_k}{(K + l - k)!} \frac{d^{K+l-k}}{dz^{K+l-k}} \left[\frac{1}{B(z)} \right]_{z=0}. \tag{4.79}$$

Collecting the terms (4.78) and (4.79) we obtain, for $l \leq -1$,

$$h_l = \sum_{n=1}^{N} \frac{Y_n \xi_n}{z_n^{K+l+1}} - \sum_{k=0}^{K-1} \theta(K + l - k) g_k \beta_{kl}, \tag{4.80}$$

where we denoted

$$Y_n = \frac{1 - z_n^2}{B_n(z_n)},$$

$$\beta_{kl} = \frac{1}{(K + l - k)!} \frac{d^{K+l-k}}{dz^{K+l-k}} \left[\frac{1}{B(z)} \right]_{z=0}. \tag{4.81}$$

Using (4.80) it is easy to calculate the sum required in (4.77). Due to the θ function, only the values $l \geq k - K$ in the second term of h_l give a non-vanishing contribution. The result is written in a compact form as:

$$\delta_2^2 = \sum_{m,n=1}^{N} A_{mn} \xi_n \xi_m + \sum_{j,k=0}^{K-1} B_{jk} g_j g_k + 2 \sum_{n=1}^{N} \sum_{k=0}^{K-1} C_{kn} g_k \xi_n \qquad (4.82)$$

where we defined

$$A_{mn} = \frac{Y_m Y_n}{z_m^K z_n^K} \frac{1}{1 - z_m z_n}, \qquad B_{jk} = \sum_{l=L}^{-1} \beta_{jl} \beta_{kl}, \qquad C_{kn} = \frac{Y_n}{z_n^K} \sum_{l=k-K}^{-1} \frac{\beta_{kl}}{z_n^{l+1}}, \qquad (4.83)$$

with $L = \max(k - K, \, j - K)$. Inserting (4.82) into the inequality (4.54) gives the allowed domain of the input values appearing in (4.53):

$$\sum_{m,n=1}^{N} A_{mn} \xi_n \xi_m + \sum_{j,k=0}^{K-1} B_{jk} g_j g_k + 2 \sum_{n=1}^{N} \sum_{k=0}^{K-1} C_{kn} g_k \xi_n \leq 1. \qquad (4.84)$$

The left hand side of this inequality is a quadratic convex expression of the parameters g_k and ξ_n, defining a convex allowed domain for these parameters.

Although the conditions (4.62) and (4.84) look quite differently, it can be checked numerically that the domains defined by them are equivalent. In particular, the case of no conditions at $z = 0$ is obtained formally by keeping only the first term in (4.84) and by setting $K = 0$ in the expression (4.83) of A_{mn}. For $N = 1$ this gives the same condition (4.64) derived above in a different way. On the other hand, the case when no conditions at $z_n \neq 0$ are imposed is obtained by keeping only the second term in (4.84) and by noting that from (4.81) and (4.83) for $N = 0$ we obtain $B_{jk} = \delta_{jk}$. Then we recover from (4.84) the same condition (4.66) derived in a different way.

4.4.3 Phase Given on a Part of the Boundary

As we discussed in Sect. 2.2, in some situations the phase of a form factor $F(t)$ is known along a part of the boundary, so we can write

$$\arg F(t + i\varepsilon) = \delta(t), \quad t_+ \leq t \leq t_{in}, \qquad (4.85)$$

where $\delta(t)$ and t_{in} are known. In this section, we consider the problem of finding the most general domain of values of $F(t)$ and its derivatives as some points inside the analyticity domain, consistent with the boundary conditions (4.6) and (4.85).

We first express the constraint (4.85) in terms of the function g defined in (4.48). It is convenient to introduce the Omnès function [11]

$$\mathscr{O}(t) = \exp\left(\frac{t}{\pi}\int_{t_+}^{\infty} dt\, \frac{\delta(t')}{t'(t'-t)}\right), \tag{4.86}$$

which we encountered already in Sect. 1.5. Here $\delta(t)$ is the known function appearing in (4.85) for $t \le t_{in}$ and is an arbitrary function, sufficiently smooth, more exactly Lipschitz continuous[5] for $t > t_{in}$. From (4.85) and (4.86) it follows that

$$\text{Im}\left[\frac{F(t+i\varepsilon)}{\mathscr{O}(t+i\varepsilon)}\right] = 0, \qquad t_+ \le t \le t_{in}. \tag{4.87}$$

Expressed in terms of the function $g(z)$ defined in (4.48) this condition becomes

$$\text{Im}\left[\frac{g(e^{i\theta})}{W(\theta)}\right] = 0, \qquad \theta \in (-\theta_{in}, \theta_{in}). \tag{4.88}$$

Here θ_{in} is defined by $\tilde{z}(t_{in}, t_0) = \exp(i\theta_{in})$ where $\tilde{z}$ is the conformal mapping (4.1), and the function $W(\theta)$ is defined as:

$$W(\theta) = \phi(e^{i\theta})O(e^{i\theta}), \tag{4.89}$$

where $\phi(z)$ is the outer function defined in Sect. 4.2, and

$$O(z) = \mathscr{O}(\tilde{t}(z)) \tag{4.90}$$

is the Omnès function expressed in the variable z.

With these notations, the problem which we must solve is to find the allowed domain taken by the values $\{g_k, \xi_n\}$ defined in (4.53), consistent with the boundary conditions (4.50) and (4.88). As above, it can be shown that this domain is given by the inequality

$$\delta_2^2 \equiv \min_{g \in \mathscr{G}'} \|g\|_2^2 \le 1, \tag{4.91}$$

where the minimum is taken over the class $\mathscr{G}'$ of analytic functions in H^2 which satisfy the conditions (4.53) and (4.88).

The constraint (4.88) can be imposed by means of a generalized Lagrange multiplier, of the type discussed in Sect. 3.8. The constraints at interior points can be treated either with Lagrange multipliers as in Sect. 4.4.1, or by the explicit implementation as in Sect. 4.4.2. For convenience, we shall adopt below the first approach.

According to the general theory presented in Sect. 3.8, the Lagrange multiplier for an abstract constraint like (4.88) belongs to the dual space, i.e. it is a functional defined on the space of functions appearing in the left hand side of this constraint. Using the Riesz representation (3.38), we can express this functional in terms of a

---

[5]Lipschitz continuity is a strong form of uniform continuity for functions, implying for instance the existence of a constant K such that $|F(z_2) - F(z_1)| \le K|z_2 - z_1|$ for arbitrary points z_i [2].

function $\lambda(\theta)$ defined on the interval $(-\theta_{in}, \theta_{in})$. Adding the terms related to the constraints at interior points, we write the Lagrangian of the minimization problem (4.91) with the constraints (4.53) and (4.88) as

$$\mathcal{L} = \sum_{k=0}^{\infty} g_k^2 + 2 \sum_{n=1}^{N} \alpha_n \left(\xi_n - \sum_{k=0}^{\infty} g_k z^k \right) \tag{4.92}$$

$$+ \frac{2}{\pi} \sum_{k=0}^{\infty} g_k \lim_{r \to 1} \int_{-\theta_{in}}^{\theta_{in}} \lambda(\theta') |W(\theta')| \operatorname{Im} \left[[W(\theta')]^{-1} r^k e^{ik\theta'} \right] d\theta'.$$

Since the function appearing in (4.88) is an odd function of θ, we can assume without loss of generality that $\lambda(\theta)$ is an odd function, $\lambda(-\theta) = -\lambda(\theta)$. The factor $|W(\theta)|$ was introduced in the integral for convenience.

For obtaining (4.92), we used the Taylor expansion (4.51) of the analytic function $g(z)$, expressing the Lagrangian in terms of the real coefficients g_k. As in Sect. 4.4.1, since the first K coefficients g_k are fixed by the constraint (4.53), we have to minimize $\mathcal{L}$ with respect to the free parameters g_k for $k \geq K$. The Lagrange multipliers $\lambda(\theta)$ and α_n are then found in the standard way by imposing the constraints, which amount to the "allignment condition", i.e. the vanishing of the correspinding terms in the Lagrangian.

In order to write the equations in a simple form, it is convenient to define the phase $\Phi(\theta)$ of the function $W(\theta)$ by

$$W(\theta) = |W(\theta)| e^{i\Phi(\theta)}, \qquad \theta \in (-\theta_{in}, \theta_{in}). \tag{4.93}$$

From (4.89) we have

$$\Phi(\theta) = \arg \phi(e^{i\theta}) + \delta(\tilde{t}(e^{i\theta}, t_0)), \quad \theta \in (-\theta_{in}, \theta_{in}), \tag{4.94}$$

where the first term is the phase of the outer function $\phi(e^{i\theta})$ and the second is the known function $\delta(t)$ written in the variable z by means of the conformal mapping (4.2). It is convenient to introduce also the functions β_n for $n = 1, \ldots, N$, by

$$\beta_n(\theta) = z_n^K \frac{\sin[K\theta - \Phi(\theta)] - \sin[(K-1)\theta - \Phi(\theta)]}{1 + z_n^2 - 2z_n \cos \theta}. \tag{4.95}$$

Then we obtain a system of coupled equations for the Lagrange multipliers $\lambda(\theta)$ and α_n. The function $\lambda(\theta)$ satisfies an integral equation of the form:

$$\sum_{k=0}^{K-1} g_k \sin[k\theta - \Phi(\theta)] = \lambda(\theta) - \sum_{n=1}^{N} \alpha_n \beta_n(\theta) - \frac{1}{2\pi} \int_{-\theta_{in}}^{\theta_{in}} d\theta' \lambda(\theta')) \mathcal{H}_\Phi(\theta, \theta'),$$

$$\tag{4.96}$$

where $\theta \in (-\theta_{in}, \theta_{in})$ and the integral kernel is

$$\mathscr{K}_\Phi(\theta, \theta') \equiv \frac{\sin[(K - 1/2)(\theta - \theta') - \Phi(\theta) + \Phi(\theta')]}{\sin[(\theta - \theta')/2]}. \tag{4.97}$$

The integral equation (4.96) is of Fredholm type if the phase $\Phi(\theta)$ is Lipschitz continuous. The remaining equations

$$-\frac{1}{\pi} \int_{-\theta_{in}}^{\theta_{in}} \lambda(\theta)\beta_n(\theta)\,d\theta + \sum_{m=1}^{N} \alpha_m \frac{(z_n z_m)^K}{1 - z_n z_m} = \bar{\xi}_n, \quad n = 1, \dots, N, \tag{4.98}$$

are algebraic and involve the solution $\lambda(\theta)$ of (4.96).

Finally, the inequality (4.91) takes the form:

$$\frac{1}{\pi} \sum_{k=0}^{K-1} g_k \int_{-\theta_{in}}^{\theta_{in}} d\theta \lambda(\theta) \sin[k\theta - \Phi(\theta)] + \sum_{n=1}^{N} \alpha_n \bar{\xi}_n \leq \bar{I}, \tag{4.99}$$

with $\bar{I}$ defined in (4.61).

As in Sect. 4.4.1, the consistency of the equations (4.98) and (4.99), viewed as a system of $N + 1$ linear relations for the N unknown parameters α_n, can be written as a positivity of a determinant which generalizes (4.62), and which describes the allowed domain of the parameters $g_0, g_1, \dots, g_{K-1}$ and $\bar{\xi}_n$. If the phase constraint is removed, one recovers condition (4.62). On the other hand, if t_{in} is increased, the allowed domain becomes smaller. The reason is that by increasing t_{in} the class of functions entering the minimization (4.91) becomes gradually smaller, leading to a larger value for minimum δ_2^2 entering the definition (4.91) of the allowed domain.

Particular forms of the above equations have been derived in [9, 12–14]. The general case of an arbitrary number of constraints at points inside the analyticity domain, presented above, has been treated for the first time in [10].

4.4.4 Phase and Modulus Given on Different Parts of the Boundary

As shown in Sect. 2.2, in some physical cases one is faced with a situation when the phase of a form factor is known on a part of the boundary:

$$\arg F(t + i\varepsilon) = \delta(t), \quad t_+ \leq t \leq t_{in}, \tag{4.100}$$

and the modulus satisfies an integral condition along the remaining part of the boundary:

$$\frac{1}{\pi} \int_{t_{in}}^{\infty} w(t)|F(t)|^2 dt \leq 1.$$ (4.101)

The problem considered in this section is to derive optimal constraints on the values and derivatives of $F(t)$ inside the holomorphy domain, in particular for real t below t_+, using as input the boundary conditions (4.100) and (4.101).

We shall use a technique proposed in [9], which allows to bring the problem to a standard form. The first step is to introduce a new function $h(t)$, by writing

$$F(t) = \mathcal{O}(t)h(t),$$ (4.102)

where $\mathcal{O}(t)$ is the Omnès function defined in (4.86). We recall that the input $\delta(t)$ in (4.86) is the known function (4.100) for $t \leq t_{in}$ and is an arbitrary smooth (Lipschitz continuous) function above t_{in}.

From (4.100) and (4.86) it follows that $h(t)$ is real on the real axis below t_{in}, since the phase of $F(t)$ is exactly compensated by the phase of $\mathcal{O}(t)$. Taking into account the fact that $h(t)$ satisfies the Schwarz reflection property, this implies that it is holomorphic on the real axis below t_{in}, having a branch cut only for $t \geq t_{in}$.

In terms of $h(t)$, inequality (4.101) becomes

$$\frac{1}{\pi} \int_{t_{in}}^{\infty} w(t)|\mathcal{O}(t)|^2 |h(t)|^2 \, dt \leq 1.$$ (4.103)

In order to bring it to a canonical form, we first perform the conformal transformation

$$\tilde{z}(t, t_0) = \frac{\sqrt{t_{in} - t_0} - \sqrt{t_{in} - t}}{\sqrt{t_{in} - t_0} + \sqrt{t_{in} - t}},$$ (4.104)

which maps the t plane cut for $t \geq t_{in}$ onto the unit disk $|z| < 1$. Note that (4.104) is obtained from (4.1) by replacing t_+ with t_{in}.

As discussed in Sect. 4.2, in order to obtain optimal constraints we must express the factors multiplying $|h(t)|^2$ in the integral (4.103) as the modulus squared of an outer function. It turns out that it is convenient to express it as a product of two outer functions. The first one, similar to the function $\phi(z)$ which we already encountered, has the modulus equal to $\sqrt{w(t)} |dt/d\tilde{z}(t, t_0)|$, and in many cases can be constructed explicitly as shown in Sect. 4.2. In addition, we define a second outer function, $\omega(z)$, which has the modulus on the cut $t \geq t_{in}$ equal to $|\mathcal{O}(t)|$, and can be calculated by the standard integral representation (3.15) given in Sect. 3.2, or using the equivalent representation in the t variable

$$\omega(z) = \exp\left(\frac{\sqrt{t_{in} - \tilde{t}(z, t_0)}}{\pi} \int_{t_{in}}^{\infty} \frac{\ln|\mathcal{O}(t')| \, dt'}{\sqrt{t' - t_{in}} \, (t' - \tilde{t}(z, t_0))} \right),$$ (4.105)

where $\tilde{t}(z, t_0)$ is the inverse of (4.104). If we define now the function $g(z)$ by

$$g(z) = F(t)\mathcal{O}(t)^{-1}\phi(z)\,\omega(z)\,, \quad t = \tilde{t}(z, t_0), \tag{4.106}$$

the condition (4.103) is written with no loss of information as

$$\frac{1}{2\pi} \int_0^{2\pi} d\theta |g(e^{i\theta})|^2 \le 1, \tag{4.107}$$

which has the standard form (4.49). Using the techniques presented in Sect. 4.4.1, we obtain from (4.107) optimal constraints on the values and derivatives of $g(z)$ (written for instance as (4.62)). By means of (4.106), these constraints are expressed finally in terms of the physical form factor $F(t)$.

One can prove [9, 10], that the bounds do not depend on the phase of the function $F(t)$ for $t > t_{in}$: the dependence of the Omnès function (4.86) on the arbitrary phase $\delta(t)$ for $t > t_{in}$ is compensated exactly by the corresponding dependence of the outer function (4.164). Furthermore, as remarked below (4.63), for a fixed weight $w(t)$ in (4.101), the bounds become stronger (weaker) if the upper bound is decreased below 1 (increased above 1), respectively.

4.5   Minimum Distance in L^2 Norm

In the previous section we considered the minimum norm problems (4.54) or (4.91), which turned out to be useful for solving the interpolation problem in L^2 norm, i.e. for describing the allowed domain of the values (4.53) consistent with the condition (4.50). In this section we shall consider a more general extremal problem, which requires to find the minimum distance

$$\delta_2^2 = \min_{g \in \mathscr{G}} \|g - h\|_2^2, \tag{4.108}$$

between a given function $h \notin H^2$ and the class $\mathscr{G} \subset H^2$ of analytic functions with prescribed values defined in (4.53). For completeness, we shall treat both cases of standard and weighted L^2 norms.

4.5.1   Standard L^2 Norm

For simplicity, we shall treat the pure Pick-Nevanlinna interpolation, when no conditions on the higher derivatives of g are imposed. As in Sect. 4.4.2, we write the general representation of the function $g(z)$ which satisfies $g(z_n) = \xi_n$ as

$$g(z) = \sum_{n=1}^{N} \xi_n \frac{B_n(z)}{B_n(z_n)} - B(z)f(z), \tag{4.109}$$

where the functions $B_n(z)$ and $B(z)$ are defined in (4.68) and the function $f(z) \in H^2$ is free of constraints.

Expressed in terms of f, the minimum distance problem (4.108) becomes

$$\delta_2 = \min_{f \in H^2} \| f - \bar{h} \|_2, \tag{4.110}$$

where

$$\bar{h}(\varsigma) = \frac{1}{B(\varsigma)} \sum_{n=1}^{N} \xi_n \frac{B_n(\varsigma)}{B_n(z_n)} - \frac{h(\varsigma)}{B(\varsigma)}. \tag{4.111}$$

We can solve (4.110) as in Sect. 4.4.2 by computing the negative-frequency Fourier coefficients of the function $\bar{h}$ and by inserting them in the sum (4.77). Alternatively, we shall resort to the duality theorem (3.44) given in Sect. 3.6, as an illustration of this powerful tool which will be used also in Sect. 4.6.

Using (3.44) and (3.41) and recalling that H^2 coincides with its dual, (3.44) allows us to write

$$\min_{f \in H^2} \| f - \bar{h} \|_2 = \sup_{F \in S^2} \left| \frac{1}{2\pi i} \oint_{|\varsigma|=1} F(\varsigma)\bar{h}(\varsigma)d\varsigma \right|, \tag{4.112}$$

where S^2 denotes the unit sphere of the Hilbert space H^2, i.e. the set of functions $F \in H^2$ which satisfy the condition $\|F\|_2 \le 1$.

Having in view the physical applications to be discussed later, we shall take the function h of the form

$$h(\varsigma) = \frac{1}{\pi} \int_{x_0}^{1} \frac{\sigma(x)}{x - \varsigma} dx, \tag{4.113}$$

where $x_0 < 1$ and $\sigma(x)$ is a known real function.

The contribution to the integral (4.112) of the first term of $\bar{h}(z)$ defined (4.111) is evaluated simply by residue theorem, which gives

$$\sum_{n=1}^{N} F(z_n) \xi_n \frac{1 - z_n^2}{B_n(z_n)}. \tag{4.114}$$

For the second term in (4.111), we apply Cauchy integral theorem along the contour shown in Fig. 4.3, which allows to pass from the integral (4.112) along the circle to an integral along the contour C which surrounds the cut of $\bar{h}(z)$. In doing so, we must first pick up the contribution of the poles of $1/B(z)$ situated below x_0 (see Fig. 4.3), which give

$$- \sum_{z_n < x_0} F(z_n) \frac{1 - z_n^2}{B_n(z_n)} \frac{1}{\pi} \int_{x_0}^{1} \frac{\sigma(x)}{x - z_n} dx. \tag{4.115}$$

We write then the integral on the contour C as

Fig. 4.3 Contour of the
Cauchy integral used for the
calculation of the integral
(4.112). The poles of $\tilde{h}(z)$
are indicated as points

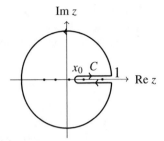

$$\frac{1}{2\pi i}\int_{x_0}^{1}\left(\frac{F(x)h(x+i\varepsilon)}{B(x+i\varepsilon)}-\frac{F(x)h(x-i\varepsilon)}{B(x-i\varepsilon)}\right)dx, \tag{4.116}$$

and use the Plemelj relations (1.29), which give

$$h(x\pm i\varepsilon)=\frac{P}{\pi}\int_{0}^{1}\frac{\sigma(y)}{y-x}dy\mp i\sigma(y), \tag{4.117}$$

where P denotes the Cauchy principal part. By applying again residue theorem for
the poles of $1/B(z)$ situated inside the contour C (see Fig. 4.3), we finally obtain the
contribution

$$-\sum_{z_n>x_0}F(z_n)\frac{1-z_n^2}{B_n(z_n)}\frac{P}{\pi}\int_{x_0}^{1}\frac{\sigma(x)}{x-z_n}dx-\frac{P}{\pi}\int_{x_0}^{1}\frac{F(x)\sigma(x)}{B(z)}dx. \tag{4.118}$$

Collecting the terms (4.114), (4.115) and (4.118) and expanding the function $F\in H^2$
as

$$F(z)=\sum_{k=0}F_k z^k, \tag{4.119}$$

we obtain from (4.110) and (4.112)

$$\delta_2=\sup_{\sum_{k=0}^{\infty}F_k^2\le 1}\left|\sum_{k=0}^{\infty}F_k c_k\right|, \tag{4.120}$$

where the coefficients c_k are written in a compact form as

$$c_k=\sum_{n=1}^{N}z_n^k\bar{Y}_n-\frac{P}{\pi}\int_{x_0}^{1}\frac{x^k\sigma(x)}{B(x)}dx,\quad k\ge 0, \tag{4.121}$$

in terms of the quantities

$$\bar{Y}_n = Y_n \left(\xi_n - \frac{P}{\pi} \int_{x_0}^1 \frac{\sigma(x)}{x - z_n} dx \right).$$ (4.122)

We used here the definition (4.81) of Y_n. For convenience, we collected in the same term the contribution of all poles, with the remark that for $z_n < x_0$ the principal part in (4.122) is superfluous.

The supremum (4.120) is obtained immediately by applying Cauchy-Schwarz inequality

$$\left| \sum_{k=0}^\infty F_k c_k \right| \leq \left[\sum_{k=0}^\infty F_k^2 \right]^{1/2} \left[\sum_{k=0}^\infty c_k^2 \right]^{1/2} \leq \left[\sum_{k=0}^\infty c_k^2 \right]^{1/2},$$ (4.123)

where in the last step we used the constraint $F \in S^2$. Since the above inequality is saturated (for $c_k = F_k$), it follows that we have

$$\delta_2^2 = \sum_{k=0}^\infty c_k^2.$$ (4.124)

This result coincides actually with (4.77): indeed, by inserting (4.119) in (4.112), we see that $c_k = \bar{h}_{-k-1}$, where $\bar{h}_l$ with $l \leq -1$ are the negative-frequency Fourier coefficients of the function $\bar{h}(\zeta)$ defined in (4.74). The duality theorem gives in this case the same result as the straightforward derivation leading to (4.77).

Using the expression (4.121) of c_k, the summation in (4.124) can be performed exactly, leading to the compact expression

$$\delta_2^2 = \sum_{m,n=1}^N \frac{\bar{Y}_n \bar{Y}_m}{1 - z_m z_n} - 2 \sum_{n=1}^N \bar{Y}_n \frac{P}{\pi} \int_{x_0}^1 \frac{\sigma(x)}{B(x)} \frac{dx}{1 - x z_n}$$
$$+ \frac{P}{\pi^2} \int_{x_0}^1 \int_{x_0}^1 \frac{\sigma(x)\sigma(y)}{B(x)B(y)} \frac{dx dy}{1 - xy}.$$ (4.125)

4.5.2 Weighted L^2 Norm

For completeness, we consider also the minimum distance problem

$$\delta_{2,w}^2 = \min_{g \in \mathscr{G}} \| g - h \|_{2,w}^2,$$ (4.126)

where the weighted L^2 norm is defined as

$$\| f \|_{2,w} \equiv \left(\frac{1}{2\pi} \int_0^{2\pi} w(e^{i\theta}) |f(e^{i\theta})|^2 d\theta \right)^{1/2},$$ (4.127)

in terms of a weight $w(\zeta) \geq 0$.

After introducing the outer function $\phi(z)$ defined by

$$w(\zeta) = |\phi(\zeta)|^2, \quad |\zeta| = 1, \tag{4.128}$$

the problem (4.126) becomes

$$\delta^2_{2,w} = \min_{g \in \mathcal{G}} \|\phi(g - h)\|^2_2 \tag{4.129}$$

and can be viewed as a standard L^2 minimum distance from the class of functions $\phi\mathcal{G} \subset H^2$ to the known function $\phi h \notin H^2$. Therefore, the solution can be found immediately from the previous one, replacing ξ_n by $\phi(z_n)\xi_n$ and h by ϕh. In particular, for h of the form (4.113), we obtain from (4.125) the compact expression

$$\delta^2_{2,w} = \sum_{m,n=1}^{N} \frac{\hat{Y}_n \hat{Y}_m}{1 - z_m z_n} - 2 \sum_{n=1}^{N} \hat{Y}_n \frac{P}{\pi} \int_{x_0}^{1} \frac{\phi(x)\sigma(x)}{B(x)} \frac{dx}{1 - xz_n}$$
$$+ \frac{P}{\pi^2} \int_{x_0}^{1} \int_{x_0}^{1} \frac{\phi(x)\phi(y)\sigma(x)\sigma(y)}{B(x)B(y)} \frac{dxdy}{1 - xy}, \tag{4.130}$$

where

$$\hat{Y}_n = Y_n \left(\phi(z_n)\xi_n - \frac{P}{\pi} \int_{x_0}^{1} \frac{\phi(x)\sigma(x)}{x - z_n} dx \right) \tag{4.131}$$

with Y_n defined in (4.81).

The solution $\delta^2_{2,w}$ can be used either for finding the minimum distance to a given function h for input value of $g(x_n)$, or for describing the allowed domain of $g(x_n)$ when the functional distance is known. In the latter case, suppose that a physical problem can be expressed in the form

$$\|g - h\|^2_{2,w} \leq 1, \tag{4.132}$$

where the function h is given. Then by repeating the arguments given below Eq. (4.54), it can be shown that the allowed domain of the parameters $g(z_n)$ is described by the inequality

$$\delta^2_{2,w} \leq 1. \tag{4.133}$$

4.6 Minimum Distance in L^∞ Norm

As we shall discuss below in Sect. 4.10, some problems in hadron physics and perturbative QCD require the solution of a functional optimization in L^∞ norm. A typical example is the quantity

$$\delta_\infty = \min_{g \in H^\infty} \|g - h\|_\infty, \tag{4.134}$$

which represents the minimum distance, measured in L^∞ norm, from a complex function $h \notin H^\infty$, given on the boundary of the unit disk, to the space H^∞ of functions analytic and bounded in $|z| < 1$.

As in Sect. 4.5.1, for solving this problem we apply the duality theorem given in Sect. 3.6. Using (3.44) and (3.41), we have

$$\min_{g \in H^\infty} \|g - h\|_\infty = \sup_{G \in S^1} \left| \frac{1}{2\pi i} \oint_{|\zeta|=1} G(\zeta) h(\zeta) d\zeta \right|, \tag{4.135}$$

where S^1 denotes the unit sphere of the Banach space H^1, i.e. the set of functions $G \in H^1$ which satisfy the condition $\|G\|_1 \leq 1$.

The equality (4.135) is automatically satisfied if h is the boundary value of an analytic function in the unit disk, since in this case the minimum norm on the left-hand side is zero, and the right-hand side of (4.135) vanishes too, by Cauchy theorem. The nontrivial case corresponds to a function h which is not the boundary value of a function analytic in $|z| < 1$ and which admits a general Fourier expansion (4.73) containing both positive and negative-frequency terms. As in the case of the distance measured in L^2 norm, we expect the minimum norm in (4.135) to depend explicitly only on the nonanalytic part, i.e. on the coefficients h_l with $l \leq -1$.

In order to evaluate the supremum on the right-hand side of (4.135), it is convenient to use a factorization theorem (given in the proof of Theorem 3.15 of Ref. [2]), according to which every function $G(z)$ belonging to the unit sphere S^1 of H^1 can be written as

$$G(z) = \phi(z) f(z), \tag{4.136}$$

where the functions $\phi(z)$ and $f(z)$ belong to the unit sphere S^2 of H^2. Therefore, if one writes the Taylor expansions

$$\phi(z) = \sum_{n=0}^{\infty} \phi_n z^n, \qquad f(z) = \sum_{m=0}^{\infty} f_m z^m, \tag{4.137}$$

the coefficients satisfy the conditions

$$\sum_{n=0}^{\infty} \phi_n^2 \leq 1, \qquad \sum_{m=0}^{\infty} f_m^2 \leq 1. \tag{4.138}$$

The duality relation (4.135) is written in terms of the functions ϕ and f as

$$\min_{g \in H^\infty} \|g - h\|_\infty = \sup_{\phi, f \in S^2} \left| \frac{1}{2\pi i} \oint_{|\zeta|=1} \phi(\zeta) f(\zeta) h(\zeta) d\zeta \right|, \tag{4.139}$$

By inserting here the expansions (4.137) we obtain, after a straightforward calculation

$$\min_{g \in H^\infty} \|g - h\|_\infty = \sup_{\{\phi_n, f_m\}} \left| \sum_{m,n=1}^\infty \mathcal{H}_{nm} \phi_{n-1} f_{m-1} \right|, \qquad (4.140)$$

where the numbers

$$\mathcal{H}_{km} \equiv h_{-k-m+1}, \quad k, m \geq 1, \qquad (4.141)$$

define a matrix $\mathcal{H}$ in terms of the negative-frequency Fourier coefficients h_l of the function h, defined in (4.74).

Matrices with elements defined as in (4.141) are called Hankel matrices [2]. If ϕ_{k-1} and $\sum_m \mathcal{H}_{km} f_{m-1}$ are viewed as the components of vectors ϕ and $\mathcal{H} f$, the absolute value of the sum in Eq. (4.140) can be written as $|\phi \cdot \mathcal{H} f|$, and the Cauchy–Schwarz inequality implies that it satisfies

$$|\phi \cdot \mathcal{H} f| \leq \|\phi\|_{L^2} \|\mathcal{H} f\|_{L^2} \leq \|\mathcal{H} f\|_{L^2}. \qquad (4.142)$$

Since the inequality (4.142) is saturated for $\phi \propto \mathcal{H} f$, it follows that the supremum in (4.140) is given by the L^2 norm of the matrix $\mathcal{H}$. The solution of the minimization problem (4.134) can then be written as

$$\delta_\infty = \|\mathcal{H}\|_{L^2} \equiv \|\mathcal{H}\|, \qquad (4.143)$$

where $\|\mathcal{H}\|$ is the spectral norm, given by the square root of the greatest eigenvalue of the positive-semidefinite matrix $\mathcal{H}^\dagger \mathcal{H}$.

In the numerical calculations, the matrix $\mathcal{H}^\dagger \mathcal{H}$ is truncated at a finite order $m = n = N$, using the fact that for large N the successive approximants tend toward the exact result (a proof of convergence is given in Appendix E of [15]). By the duality theorem, the initial functional minimization problem (4.134) was reduced to a simpler numerical computation, which can be treated with suitable algorithms (see for instance [16]).

4.7   Approximate Solution of the Minimum Distance in L^∞ Norm

We show now that it is possible to approximate the minimum distance δ_∞ in L^∞ norm by a minimum distance $\delta_{2,w}$ measured in a suitable weighted L^2 norm. We follow an argument based on the duality theorem, which was given in [6].

We start from (4.139), where the r.h.s. requires the calculation of the supremum upon two sets of functions, $\phi(z)$ and $f(z)$, analytic in $|z| < 1$ and of L^2 norm bounded by 1. In Sect. 4.6, the supremum on both functions have been evaluated simultaneously. It is convenient however to calculate first the supremum upon one class of functions, say f, keeping the other one fixed. We note that the r.h.s. of (4.139) can be written as

$$\left| \frac{1}{2\pi i} \oint_{|\zeta|=1} \phi(\zeta) f(\zeta) h(\zeta) d\zeta \right| = \left| \sum_{n=1}^{\infty} d_n f_{n-1} \right|, \tag{4.144}$$

where f_n are the Taylor coefficients defined in (4.137) and $d_n \equiv h_{-n}$ are the negative-frequency Fourier coefficients of the product $\phi(\zeta) h(\zeta)$, written as

$$d_n = \frac{1}{2\pi i} \int_{|\zeta|=1} \zeta^{n-1} \phi(\zeta) h(\zeta) d\zeta, \quad n \geq 1. \tag{4.145}$$

Then Eq. (4.139) becomes

$$\delta_\infty = \sup_{\phi \in S^2} \sup_{f \in S^2} \left| \sum_{n=1}^{\infty} d_n f_{n-1} \right|. \tag{4.146}$$

The supremum upon the coefficients f_n subject to the second condition (4.138) can be evaluated immediately using Cauchy-Schwarz inequality, as in the previous section. This leads to

$$\delta_\infty = \sup_{\phi \in S^2} \left[\sum_{n=1}^{\infty} d_n^2 \right]^{1/2}. \tag{4.147}$$

By taking into account the expression (4.145) of the coefficients d_n, we recognize in the r.h.s. of (4.147) the minimum distance

$$\delta_{2,w} = \min_{g \in H^2} \| g - h \|_{2,w}, \tag{4.148}$$

measured in the weighted L^2 norm

$$\| f \|_{2,w} \equiv \left(\frac{1}{2\pi} \int_0^{2\pi} |\phi(e^{i\theta})|^2 |f(e^{i\theta})|^2 d\theta \right)^{1/2}. \tag{4.149}$$

So, we can write

$$\delta_\infty = \sup_{\phi \in S^2} \delta_{2,w}. \tag{4.150}$$

We emphasize that this is an exact relation, which states that the minimum distance δ_∞ in the L^∞ norm is the largest $\delta_{2,w}$, for $\phi(z)$ subject to the first condition (4.138).

Of course, the problem is not yet solved, we still have to calculate the supremum in (4.150). The procedure is useful if one can find a suitable, simple parametrization of the functions ϕ, such that the maximization upon this limited class approximates well the exact δ_∞. It turns out that such a choice exists [6]: the main observation is that one can obtain approximately the maximum modulus squared of a function on a certain interval by computing the normalized integral of its modulus squared in a variable that dilates the region where the modulus of the function reaches its maximum. Therefore, one can approximate the L^∞ norm (3.9) of an arbitrary function by an L^2 norm (3.8) for $p = 2$, defined on the integration range distorted by a suitable change of variable.

In order to obtain this variable, we consider the conformal mapping of the unit disk $|z| \leq 1$ onto itself, achieved by the Blaschke transformation, which we already met several times:

$$z' = \frac{z - a}{1 - a^* z},\tag{4.151}$$

where a is an arbitrary parameter with $|a| < 1$. Since we consider real analytic functions, we can restrict a to real values in the range $(-1, 1)$. Making this change of variable in the standard L^2 norm (obtained from (4.149) for $\phi = 1$) introduces the Jacobian $|dz'/dz|$, which corresponds to a weight function $\phi(z)$ in the weighted L^2_w norm, of the form

$$\phi(z) = \frac{\sqrt{1 - a^2}}{1 - az}, \qquad a \in (-1, 1).\tag{4.152}$$

It is easy to check that this function satisfies the first condition (4.138).

Using the above remark, we can restrict the choice of admissible functions $\phi(z)$ in (4.150) to the simple form (4.152). We can then obtain with good approximation the minimum distance δ_∞ by a relatively simple algorithm: first, $\delta_{2,w}$ is calculated for a fixed choice of a in (4.152). Then the parameter a is varied in the range $(-1, 1)$ and the largest value of $\delta_{2,w}$ is retained. The procedure is convenient especially when $\delta_{2,w}$ can be written in an analytic compact form, as in (4.130). Moreover, hints on the optimal value of the parameter a can be often obtained from the physical input, making the algorithm quickly convergent.

4.8 Solution of a Problem of Analytic Extrapolation

In Sect. 1.6 we briefly discussed the instability of analytic extrapolation, which means that two analytic functions which are very close along a contour may differ by an arbitrary amount at points outside the contour. For illustration we formulated a problem which requires to find the class of functions $f(z)$ analytic in the unit disk, subject to the conditions (1.91) and (1.92). In this section we give the solution of this problem.

For convenience, we assume that the region Γ_1 is the right semicircle $\zeta = e^{i\theta}$, $-\pi/2 < \theta < \pi/2$, see Fig. 4.4. This situation can be achieved by a suitable conformal mapping of the original complex plane. We further consider for simplicity that the point z_0 is real.

Fig. 4.4 Unit disk in the
complex z-plane, with the
boundary regions Γ_1 and Γ_2

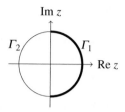

We assume as a technical step the condition

$$|f(\zeta)| \le M, \qquad \zeta \in \Gamma_2, \tag{4.153}$$

where M is a real number and Γ_2 is the remaining part of the boundary, shown in
Fig. 4.4.

Let us define now an outer function $C(z)$ such that

$$|C(\zeta)| = \begin{cases} \frac{1}{\varepsilon} & \text{for} \ \ \zeta \in \Gamma_1 \\ \frac{1}{M} & \text{for} \ \ \zeta \in \Gamma_2. \end{cases} \tag{4.154}$$

From (1.91) and (4.153) it follows that the new real analytic function

$$g(z) = f(z)C(z) \tag{4.155}$$

satisfies the upper bound

$$|g(\zeta)| \le 1, \qquad \zeta \in \Gamma, \tag{4.156}$$

where $\Gamma \equiv \Gamma_1 \cup \Gamma_2$ is the whole boundary, $|\zeta| = 1$, of the unit disk. From (1.92)
and (4.155) it follows that

$$g(z_0) = \delta_0 \, C(z_0). \tag{4.157}$$

We proceed now to construct explicitly an analytic function $g(z)$ which satisfies the
conditions (4.156) and (4.157). The construction is based on the Blaschke factor

$$B(z; z_0) = \frac{z - z_0}{1 - z_0^* z}, \qquad |z_0| < 1, \tag{4.158}$$

where $z_0 = z_0^*$. One can check that in this case $B(z; z_0)$ satisfies the Schwarz condition $B(z^*; z_0) = B^*(z; z_0)$.

We further define a new function $h(z)$ by

$$h(z) = \frac{B(g(z); g_0)}{B(z; z_0)} = \frac{g(z) - g(z_0)}{z - z_0} \frac{1 - z z_0}{1 - g(z)g(z_0)}. \tag{4.159}$$

By construction, $h(z)$ is analytic in $|z| < 1$, since the zero in the first denominator is compensated by the zero in the first numerator, and the second denominator cannot vanish. From the property $|B(\zeta; z_0)| = 1$, it follows that h satisfies the boundary constraint

$$|h(\zeta)| \le 1, \qquad |\zeta| = 1, \tag{4.160}$$

and is otherwise arbitrary, in particular there are no constraints on it at $z = z_0$.

We can take in particular $h(z) = \chi$, where χ is a constant with $|\chi| \le 1$. Then, using (4.159) and (4.155) we obtain the function $f(z)$ as

$$f(z) = \frac{1}{C(z)} \frac{\delta_0 C(z_0) + \chi B(z; z_0)}{1 + \delta_0 C(z_0) \chi B(z; z_0)}. \tag{4.161}$$

This expression satisfies the original conditions (1.91) and (1.92) for an arbitrary δ_0.

There is however a condition that must be satisfied for the above procedure to work: the constant $g(z_0)$ defined in (4.157) must satisfy the condition $|g(z_0)| < 1$ which results from (4.156) and the maximum modulus principle [2]. Let us investigate whether this condition is fulfilled. We recall that the outer function $C(z)$ is obtained from the integral representation (3.15) as

$$C(z) = \exp \left[\frac{1}{2\pi} \ln \frac{1}{\varepsilon} \int_{-\pi/2}^{\pi/2} d\theta \frac{e^{i\theta} + z}{e^{i\theta} - z} + \frac{1}{2\pi} \ln \frac{1}{M} \int_{\pi/2}^{3\pi/2} d\theta \frac{e^{i\theta} + z}{e^{i\theta} - z} \right]. \tag{4.162}$$

The explicit evaluation of the integral in (4.162) is straightforward, leading to

$$C(z) = \varepsilon^{-\omega(z)} M^{\omega(z)-1}, \tag{4.163}$$

where

$$\omega(z) = \frac{1}{2} + \frac{2}{\pi} \arctan z. \tag{4.164}$$

It is easy to check that for $z \in (-1, 1)$, this function satisfies $0 < \omega(z) < 1$.

We use now (4.157) and impose the condition $|g(z_0)| < 1$, from which we obtain the constraint

$$|\delta_0| < \varepsilon^{\omega(z_0)} M^{1-\omega(z_0)}. \tag{4.165}$$

Assuming for instance ε to be very small, the first factor in (4.165) can be made very small since $\omega(z_0) > 0$. This low value can be compensated by taking a large value of the free parameter M, which contributes also with a positive exponent $1 - \omega(z_0) > 0$. So, we can obtain an arbitrary value of δ_0 by properly adjusting the allowed upper bound on the modulus $|f(\zeta)|$ along the part Γ_2 of the boundary.

From the above solution, one can see when the procedure breaks down and the analytic continuation cannot produce arbitrary values. This happens, for instance, when the functions satisfy on Γ_2 a conditions of the form (4.153) with a known, finite M. Then from (4.165) it follows that $f(z_0)$ goes indeed to zero when $\varepsilon \to 0$, i.e. the analytic continuation is stable. The condition (4.153) is a particular example of Tikhonov regularization mentioned in Sect. 1.6, since it restricts the class $\mathscr{C}$ of analytic functions to a compact set.

By applying the construction described above to the function $h(z)$, one can implement an additional prescribed value for $f(z)$ at a second point z_1. As shown in [17] the problem can be generalized, i.e. one can construct a function $f(z)$ which has arbitrary values at a set z_i of points inside the analyticity domain. Again, the problem can be solved by properly adjusting the upper bound M supposed to hold on Γ_2. It follows that, if no information on the behavior on Γ_2 is available, the predictive power of analyticity outside the region Γ_1 is practically nil.

4.9 Optimization of Series Convergence by Conformal Mappings

We introduced in Sect. 4.1 the conformal mapping of the analyticity domain onto the unit disk for bringing a boundary value physical problem to a canonical mathematical form. As we shall discuss in this section, the conformal mappings play an important role also for improving the convergence of the series expansions used for the representation of functions of physical interest. In this context, the method of conformal mappings was introduced in particle physics in Refs. [18–20].

By this method, a series in powers of a certain variable, convergent in a disk of positive radius around the origin, is replaced by a series in powers of another variable, which actually performs the conformal mapping of the original complex plane (or a part of it) onto a disk of radius equal to unity in the transformed plane. As shown in [18, 20], the new series converges in a larger region, well beyond the disk of convergence of the original expansion, and also has an increased asymptotic convergence rate at points lying inside this disk.

The conformal mapping is actually not unique: one can map onto the unit disk the whole analyticity domain from the original complex plane[6] or only a part of it. It may intuitively seem that the larger the domain mapped onto the unit disk, the better the convergence properties of the series expansion in powers of the new variable. This is indeed true, and we shall give this hope a precise mathematical form. The result, proved in [18], important and interesting as it is for a number of applications, did not raise enough interest as it deserved, in spite of the many applications of the conformal mapping method during the last decades.

We consider for convenience a function $B(u)$ holomorphic in the complex u plane with two cuts along the real axis as shown in the left part of Fig. 4.5. The Taylor expansion

---

[6]We consider in this discussion that the analyticity domain is a simply-connected domain.

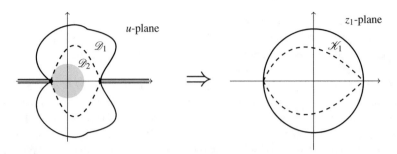

Fig. 4.5 Left: the domains $\mathscr{D}_1$ and $\mathscr{D}_2$ in the u-plane. Right: the z_1 complex plane with the unit disk $\mathscr{K}_1$ defined in Lemma 1. The domain $\mathscr{D}_2$ is mapped inside the dashed curve

$$B(u) = \sum_{n=0}^{\infty} b_n \, u^n \qquad (4.166)$$

converges only inside the circle passing through the nearest singularity of $B(u)$ (the convergence domain is shown as a grey disk in Fig. 4.5 left). The convergence domain can be increased by a conformal mapping. The following two lemmas [21], which develop the idea of the proof given in [18], show which is the variable that provides the best asymptotic rate of convergence.

Lemma 1 *Let $\mathscr{D}_1$ and $\mathscr{D}_2$ be two domains in the complex u-plane, with $\mathscr{D}_2 \subset \mathscr{D}_1$, $\mathscr{D}_2 \neq \mathscr{D}_1$ (see Fig. 4.5), such that the conformal mappings $\mathscr{D}_1 \to \mathscr{K}_1$ and $\mathscr{D}_2 \to \mathscr{K}_2$ exist, where $\mathscr{K}_1$ and $\mathscr{K}_2$ are unit disks. Denote these two mappings by*

$$z_1 = \tilde{z}_1(u) : \mathscr{D}_1 \to \mathscr{K}_1 = \{z_1 : |z_1| < 1\},$$
$$z_2 = \tilde{z}_2(u) : \mathscr{D}_2 \to \mathscr{K}_2 = \{z_2 : |z_2| < 1\}. \qquad (4.167)$$

Let Q be a point of $\mathscr{D}_2$, $Q \in \mathscr{D}_2$, such that $\tilde{z}_1(Q) = 0$ and $\tilde{z}_2(Q) = 0$. Then

$$|\tilde{z}_1(u)| < |\tilde{z}_2(u)|, \quad \text{for all } u \in \mathscr{D}_2, \ u \neq Q. \qquad (4.168)$$

Proof Let us define

$$f(z_2) = \tilde{z}_1(\tilde{z}_2^{[-1]}(z_2)) \qquad (4.169)$$

for $z_2 \in \mathscr{K}_2$, where $\tilde{z}_2^{[-1]}$ is the inverse to $\tilde{z}_2$, which exists since $\tilde{z}_2(u)$ is a conformal mapping.

The function $f(z_2)$ is holomorphic on the unit disk $\mathscr{K}_2$ of the z_2-plane and maps this disk into the unit disk $\mathscr{K}_1$ of the z_1-plane, i.e. $|f(z_2)| \leq 1$. Moreover, since $\tilde{z}_1(Q) = 0$ and $\tilde{z}_2(Q) = 0$ by assumption, it follows that $f(0) = 0$.

We now apply Schwarz lemma [22], which states that if a function $F(z)$ is holomorphic on the disk $|z| < 1$ and satisfies the conditions $F(0) = 0$ and $|F(z)| < 1$ for $|z| < 1$, then

$$|F(z)| \le |z| \tag{4.170}$$

everywhere in $|z| < 1$. Besides, if the equality sign occurs in (4.170) at least at one interior point, then it takes place everywhere and $F(z)$ has the form $F(z) = z\exp(i\alpha)$ with α real.

Applying Schwarz lemma to the function f defined in (4.169), we have $|f(z_2)| \le |z_2|$ for $z_2 \in \mathcal{K}_2$. Using the definition (4.169) and the obvious relation $\tilde{z}_2^{[-1]}(z_2) = u$ for $u \in \mathcal{D}_2$, we obtain

$$|\tilde{z}_1(u)| \le |\tilde{z}_2(u)|, \quad u \in \mathcal{D}_2. \tag{4.171}$$

Ignoring the mappings that reduce to mere rotations according to Schwarz lemma, we are left with a sharp inequality in (4.171),

$$|\tilde{z}_1(u)| < |\tilde{z}_2(u)|, \quad u \in \mathcal{D}_2, \quad u \ne Q, \tag{4.172}$$

which proves Lemma 1.

Lemma 2 *Let $\mathcal{D}_1$ and $\mathcal{D}_2$ be the domains defined in Lemma 1, $\tilde{z}_1(u)$ and $\tilde{z}_2(u)$ the mappings (4.167) and $B(u)$ a function holomorphic in $\mathcal{D}_1$. Define the expansions*

$$B(u) = \sum_{n=0}^{\infty} c_{n,1}(\tilde{z}_1(u))^n, \tag{4.173}$$

$$B(u) = \sum_{n=0}^{\infty} c_{n,2}(\tilde{z}_2(u))^n, \tag{4.174}$$

which are convergent for $z_1 \equiv \tilde{z}_1(u) \in \mathcal{K}_1$ and $z_2 \equiv \tilde{z}_2(u) \in \mathcal{K}_2$, respectively. Assume in addition that the limits $\lim_{n\to\infty} \sqrt[n]{|c_{n,1}|}$ and $\lim_{n\to\infty} \sqrt[n]{|c_{n,2}|}$ exist[7] and are equal to one:

$$\lim_{n\to\infty} \sqrt[n]{|c_{n,1}|} = \lim_{n\to\infty} \sqrt[n]{|c_{n,2}|} = 1. \tag{4.175}$$

Then a positive integer $N = N(u)$ exists such that the following inequality holds:

$$\mathcal{R}_n(u) = \left| \frac{c_{n,1}(\tilde{z}_1(u))^n}{c_{n,2}(\tilde{z}_2(u))^n} \right| < 1, \tag{4.176}$$

for any n integer, $n > N$, and $u \in \mathcal{D}_2$, $u \ne Q$.

Proof The relations (4.175) imply that, for large enough n, the coefficients $|c_{n,j}|$ can be represented in the form

---

[7]The essence of this requirement is that the expansions (4.173) and (4.174) have equal radii of convergence. This assumption is nontrivial, because the expanded function might be of such a form that certain singularities of $B(u)$ in z_1 or z_2 variables might disappear.

$$|c_{n,j}| = e^{g_j(n)}, \qquad j = 1, 2, \tag{4.177}$$

where $g_j(n)$ are real-valued functions, subject to the conditions $\lim_{n\to\infty} g_j(n)/n = 0$, $j = 1, 2$. Then, the ratio defined in (4.176) can be written as

$$\mathscr{R}_n(u) = e^{g(n)} \times (\rho(u))^n, \tag{4.178}$$

where

$$g(n) = g_1(n) - g_2(n), \qquad \rho(u) = |\tilde{z}_1(u)/\tilde{z}_2(u)|. \tag{4.179}$$

Taking the logarithm of (4.178), one obtains, for large n, the inequality

$$\ln \mathscr{R}_n(u) = n \left[\frac{g(n)}{n} + \ln \rho(u) \right] < 0, \tag{4.180}$$

since from (4.179) it follows that $\lim_{n\to\infty} g(n)/n = 0$, while $\rho(u) < 1$ for all $u \in \mathscr{D}_2$, $u \neq Q$, according to Lemma 1. This implies (4.176), proving Lemma 2.

The practical consequence is that the best asymptotic convergence rate is obtained by expanding $B(u)$ in powers of the variable $\tilde{z}(u)$ that maps the whole holomorphy domain of B, i.e. the cut u-plane shown in Fig. 4.5, onto the unit disk $|z| < 1$. This variable is known in the literature as "optimal conformal mapping" [18].

We emphasize finally that the above discussion considers the improvement of power series, in particular of the Taylor expansions at a certain point. In practical representation of data, other situations may occur. Thus, when one wishes to approximate a function along an interval, the most natural expansion is given in terms of orthonormal polynomials on that interval. As it is known [23], the domain of convergence of an expansion in orthonormal polynomials is an ellipse passing through the first singularity of the function. In order to accelerate the rate of convergence along the range of interest, one must conformally map the function's analyticity domain onto the interior of an ellipse, such that this range is mapped on the segment between the focal points [24]. If the original analyticity domain is the complex plane with two cuts along the real axis as in Fig. 4.5 left and the physical range is a real interval situation between the cuts, the optimal conformal mapping is expressed in terms of elliptic functions [24].

4.10 Brief Review of Physical Applications

The mathematical techniques presented above have been applied to various optimization problems in hadron physics, either before or after the development of the Standard Model. In this section we briefly review some of these applications, in a historical perspective, indicating in each case the specific mathematical technique used. Since in the next two chapters we shall discuss in more detail two classes of recent applications, namely the hadronic form factors and perturbative QCD, these subjects are only briefly mentioned in this section.

By using analyticity and unitarity in field theory, Meiman [25] and Okubo [26] formulated for the first time an optimization problem in L^2 norm, of the type considered in Sect. 4.4. Starting from a dispersion relation (a Källen-Lehmann representation) for a Green function in field theory and using unitarity and positivity for its spectral function on the timelike axis, they derived inequalities of the form (4.6) for the relevant form factors.[8] The L^2-norm boundary condition was then exploited with the techniques presented in Sects. 4.4.1 and 4.4.2 for deriving model-independent bounds on the values of the form factors in physically accessible regions. A mathematical improvement of the method was achieved by Micu [12] and Auberson et al. [13], who included information on the phase of the form factor along a part of the boundary and derived integral equations of the form obtained above in Sect. 4.4.3 for a generalized Lagrange multiplier. Finally, the role of the Hardy H^p spaces in the derivation of bounds on the form factors was emphasized for the first time by Raina [27].

However, the studies performed before the advent of QCD have been obstructed to a certain extent by the fact that the asymptotic behavior of the Green functions was not known from a reliable theory. In general, it was assumed that the current correlators satisfy unsubtracted dispersion relations, which led in some cases to distorted predictions of the formalism for the corresponding form factors. The correct asymptotic behavior of the field correlators could be established only after the discovery of QCD as the proper quantum field theory of strong interactions. The first application of the Meiman-Okubo approach in a modern context was made by Bourrely, Machet and de Rafael [28], who derived bounds on the weak form factors which describe the semileptonic $K \to \pi \ell \nu$ decays, using perturbative QCD to calculate the field correlators in the deep Euclidian region.

After a decade, at the beginning of the 1990s, the method was resuscitated by de Rafael and Taron [29], who used it for deriving rigorous bounds on the so-called Isgur-Wise function [30], which parametrizes the semileptonic decays of heavy mesons in the limit of infinite mass. Subject first to criticism in Refs. [31–34] because it neglected the sub-threshold singularities of the form factors, this approach was subsequently adapted to take into account in an optimal way these singularities, which opened the route to many applications to various weak semileptonic form factors. We do not present them here, since we devote Sect. 5.3 of the next chapter to this subject.

Optimization techniques in L^2 norm, like those presented above in Sect. 4.4, have been applied also to the pion electromagnetic form factor, starting from the early 1970s [7, 35]. In particular, a first rigorous lower bound on the two-pion contribution to the muon anomaly was derived in [8, 36, 37]. The inclusion of the phase on a part of the boundary, known in this case from $\pi\pi$ scattering by Fermi-Watson theorem, was achieved in [9], using the methods presented above in Sects. 4.4.3 and 4.4.4. This paper opened the way to a series of studies of the pion electromagnetic form

---

[8]We shall present in more detail this approach in Sect. 5.1.

factor by functional optimization methods, which gradually increased the precision of the predictions for this important hadronic observable. Some recent results on the pion electromagnetic form factor will be presented in Sect. 5.4.

The interest in the optimization and functional analysis methods was stimulated also by the investigation of analytic continuation. As mentioned in Sect. 1.6, in the late 1960s and early 1970s it became more and more clear that the inevitable instability of analytic continuation may affect the physical predictions in the frame of the analytic S-matrix theory. As illustrated by the problem formulated in Sect. 1.6 and solved in Sect. 4.8, the power of analyticity for the extrapolation from a limited part of the boundary to a point outside the original range is practically zero, if some quantitative information on the behavior on the remaining part of the boundary is not available. This problem is encountered, in particular, in the determination of the masses and widths of resonances, defined as poles of the scattering amplitudes on higher Riemann sheets of the complex-energy plane. The instability of the analytic extrapolation may distort the prediction of these parameters, especially for broad (short-lived) resonances associated to poles far from the physical region.[9]

Attempts to control the extrapolation of an error-affected input given on a part of the boundary of the analyticity domain have been done either by means of modified dispersion relations [38, 39], or by formulating and solving suitable minimization problems in L^2 or L^∞ norms [15, 40, 41], of the form presented above in Sects. 4.4 and 4.6. In particular, a parametrization-free method for finding resonances from error-affected data, based on the minimum distance problem in L^∞ norm presented in Sect. 4.6, has been proposed in [42]. A review of various techniques of analytic extrapolation in particle physics before 1975 can be found in [17].

Methods of functional optimization and modified dispersion relations have been applied also to the scattering amplitudes of several specific processes. As we have seen in Sect. 2.2, the study of Compton scattering on proton required the application of techniques discussed in Sect. 4.3 for writing a quadratic expression involving the analytic invariant amplitudes given on the boundary in an equivalent diagonal form. The specific Blaschke-Potapov factorization in the case of proton Compton scattering was found in [43], allowing the application, in the subsequent studies [44, 45], of the optimization methods reviewed in Sects. 4.4, 4.6 and 4.7. More recently, the optimization methods presented in Sect. 4.6 have been applied in [46] for finding optimal constraints on the subtraction functions in the dispersion relations for the forward virtual Compton scattering on proton, of interest for the evaluation of the Cottingham formula for the proton-neutron electromagnetic mass difference, the nucleon polarizabilities and the proton radius puzzle.

Techniques of functional optimization have been used also for improving the rigorous results derived from axiomatic field theory (we mentioned several rigorous results for pion-pion scattering in Sect. 1.2). In the analysis performed in [6], the

---

[9]A notorious example is the lowest $I = J = 0$ resonance in $\pi\pi$ scattering, known as σ or $f_0(500)$, which for some time was even expelled from the PDG tables due to the large uncertainties in its mass and width. A precise determination of this resonance was possible by the progress in the development of χPT and dispersion theory (Roy equations) for pion-pion scattering, mentioned in Sect. 2.2.

axiomatic constraints on the amplitude of the process $\pi^0\pi^0 \to \pi^0\pi^0$ have been shown to define an admissible functional convex set in the space H^2, allowing the application of the duality theorem for convex sets (3.50) given in Sect. 3.7. The lower bound $a^{00} > -1.70$ on the S-wave scattering length derived in this way slightly improved the previous axiomatic lower bound $a^{00} > -1.75$ derived in [47], which we quoted in Sect. 1.2.

After the development of the Standard Model, in the 1980s, techniques of functional analysis have been considered as an alternative to the standard finite-energy sum rules (FESR), proposed originally in [48] for connecting OPE and perturbative QCD to the low-energy hadronic observables. We discussed this approach for a generic polarization function $\Pi(s)$ in Sect. 2.4.2. The studies reported in the literature [49–51] have been based on the optimization problems in L^2 and L^∞ norms treated in Sects. 4.4, 4.5 and 4.6. The interest in this approach increased recently in connection with the study of the so-called quark-hadron duality violation, which we shall discuss in more detail in Chap. 6.

Besides the dispersion relations, the analytic parametrizations, in particular the expansions of the functions of interest in power series, represent another important method of implementing analyticity. In this context, it is worth mentioning that optimized expansions, based on orthonormal polynomials in conformal-mapping variables, have been proposed in [24, 52, 53] as alternatives to the standard partial-wave expansions of the scattering amplitudes.

The method of conformal mapping, briefly reviewed in Sect. 4.9, has been applied also to the perturbative expansions in field theory, in particular to perturbative QCD. It is known that the perturbative expansions in quantum field theory are in most cases divergent series, with coefficients growing factorially at large orders. It turns out actually that the method described in Sect. 4.9 is not directly applicable to the formal perturbative series of the Green functions in powers of the coupling constant, because these functions are singular at the point of expansion. A solution was proposed in [54, 55] by the so-called "order-dependent" conformal mappings, where the singularity is shifted away from the origin by a certain amount at each finite order, and tends to the origin only when an infinite number of terms are considered. On the other hand, the method of conformal mappings can be applied in a straightforward way to the Borel transforms of the Green functions, which are holomorphic in a region containing the origin $u = 0$ of the Borel complex plane. This approach will be discussed in more detail in Sect. 6.3.

References

1. Z. Nehari, *Conformal Mapping* (McGraw-Hill, New York, 1952)
2. P.L. Duren, *Theory of H^p Spaces* (Academic Press, New York, 1970)
3. D.G. Luenberger, *Optimization by Vector Space Methods* (Wiley, New York, 1968)
4. H. Kober, *Dictionary of Conformal Representations* (Dover, New York, 1957)
5. V.P. Potapov, Trudy Moskov. Mat. Obsc. **4**, 125 (1955). (in Russian)
6. I. Caprini, P. Dita, J. Phys. A **13**, 1265 (1980)

7. V. Singh, A.K. Raina, Fortsch. Phys. **27**, 561 (1979)
8. A.K. Raina, V. Singh, J. Phys. G **3**, 315 (1977)
9. I. Caprini, Eur. Phys. J. C **13**, 471 (2000)
10. G. Abbas, B. Ananthanarayan, I. Caprini, I. Sentitemsu Imsong, S. Ramanan, Eur. Phys. J. A **45**, 389 (2010)
11. R. Omnès, Nuovo Cim. **8**, 316 (1958)
12. M. Micu, Nucl. Phys. B **44**, 531 (1972); Phys. Rev. D **7**, 2136 (1973)
13. G. Auberson, G. Mahoux, F.R.A. Simão, Nucl. Phys. B **98**, 204 (1975)
14. C. Bourrely, I. Caprini, Nucl. Phys. B **722**, 149 (2005)
15. S. Ciulli, G. Nenciu, J. Math. Phys. **14**, 1675 (1973)
16. I. Caprini, M. Săraru, C. Pomponiu, M. Ciulli, S. Ciulli, I. Sabba-Stefanescu, Comput. Phys. Commun. **18**, 305 (1979)
17. S. Ciulli, C. Pomponiu, I. Sabba-Stefanescu, Phys. Rept. **17**, 133 (1975)
18. S. Ciulli, J. Fischer, Nucl. Phys. **24**, 465 (1961)
19. W.R. Frazer, Phys. Rev. **123**, 2180 (1961)
20. I. Ciulli, S. Ciulli, J. Fischer, Nuovo Cim. **23**, 1129 (1962)
21. I. Caprini, J. Fischer, Phys. Rev. D **84**, 054019 (2011)
22. W. Rudin, *Real and Complex Analysis* (McGraw-Hill, New York, 1966)
23. J.L. Walsh, *Interpolation and Approximation by Rational Functions in the Complex Domain* (American Mathematical Society, Providence, RI, 1956)
24. R.E. Cutkoski, B.B. Deo, Phys. Rev. **174**, 1859 (1968)
25. N. N. Meiman, Zh. Eksp. Teor. Fiz. **44**, 1228 (1963) [Sov. Phys. JETP **17**, 830 (1963)]
26. S. Okubo, Phys. Rev. D **3**, 2807 (1971); Phys. Rev. D **4**, 725 (1971)
27. A.K. Raina, Lett. Math. Phys. **2**, 513 (1978)
28. C. Bourrely, B. Machet, E. de Rafael, Nucl. Phys. B **189**, 157 (1981)
29. E. de Rafael, J. Taron, Phys. Lett. B **282**, 215 (1992)
30. N. Isgur, M.B. Wise, Phys. Lett. B **237**, 527 (1990)
31. A.F. Falk, M.E. Luke, M.B. Wise, Phys. Lett. B **299**, 123 (1993)
32. B. Grinstein, P.F. Mende, Phys. Lett. B **299**, 127 (1993)
33. C.E. Carlson, J. Milana, N. Isgur, T. Mannel, W. Roberts, Phys. Lett. B **299**, 133 (1993)
34. J.G. Korner, D. Pirjol, C. Dominguez, Phys. Lett. B **301**, 257 (1993)
35. I. Raszillier, Commun. Math. Phys. **26**, 121 (1972)
36. G. Nenciu, I. Raszillier, Nuovo Cim. A **11**, 319 (1972)
37. A.K. Raina, V. Singh, Nucl. Phys. B **139**, 341 (1978)
38. S. Ciulli, J. Fischer, Nucl. Phys. B **24**, 537 (1970)
39. J. Fischer, J. Pišút, P. Prešnajder, J. Sebesta, Czech. J. Phys. **19**, 1486 (1969)
40. J. Pišút, P. Prešnajder, Nucl. Phys. B **12**, 110 (1969)
41. P. Prešnajder, J. Pišút, Nuovo Cim. A **3**, 603 (1971)
42. I. Caprini, S. Ciulli, A. Pomponiu, I. Sabba-Stefanescu, Phys. Rev. D **5**, 1658 (1972)
43. I. Guiasu, I. Raszillier, E.E. Radescu, Ann. Phys. **127**, 436 (1980)
44. I. Caprini, I. Guiasu, E.E. Radescu, Phys. Rev. D **25**, 1808 (1982)
45. I. Caprini, Phys. Rev. D **27**, 1479 (1983)
46. I. Caprini, Phys. Rev. D **93**, 076002 (2016)
47. C. Lopez, G. Mennessier, Phys. Lett. **58B**, 437 (1975)
48. M.A. Shifman, A.I. Vainshtein, V.I. Zakharov, Nucl. Phys. B **147**, 385–448 (1979)
49. I. Caprini, C. Verzegnassi, Nuovo Cim. A **90**, 388 (1985)
50. G. Auberson, G. Mennessier, Commun. Math. Phys. **121**, 49 (1989)
51. I. Caprini, Phys. Rev. D **44**, 1569 (1991)
52. R.E. Cutkosky, Ann. Phys. **54**, 350 (1969)
53. S. Ciulli, Nuovo Cim. A **61**, 787 (1969); Nuovo Cim. A**62**, 301 (1969)
54. R. Seznec, J. Zinn-Justin, J. Math. Phys. **20**, 1398 (1979)
55. J. Zinn-Justin, U.D. Jentschura, J. Math. Phys. **51**, 072106 (2010)

Chapter 5
Constraints on Hadronic Form Factors

In this chapter, we discuss several applications which illustrate the usefulness of the functional analysis and optimization methods for improving the knowledge of the weak and electromagnetic hadronic form factors. We first present the method of "unitarity bounds", proposed in the early 1970s by Meiman and Okubo for deriving model-independent bounds on the semileptonic form factors. The development of the method in the frame of the Standard Model is then reviewed, emphasizing the increased strength of the formalism when it is combined with additional theoretical information provided by heavy-quark symmetry, chiral perturbation theory or lattice QCD. Finally, we show how this approach leads to precise predictions for the pion electromagnetic form factor, in particular for the charge radius of the pion. We briefly describe also the way in which the rigorous bounds can be merged with statistical simulations.

5.1 Meiman-Okubo Unitarity Bounds

As already mentioned in Sect. 4.10, Meiman [1] and Okubo [2] formulated for the first time an optimization problem in L^2 norm, by exploiting analyticity and unitarity in field theory. Starting from a dispersion relation (a Källen–Lehmann representation) for a suitable Green function and using unitarity and positivity for its spectral function on the timelike axis, they derived inequalities of the form (4.6) for the form factors parametrizing the lowest contributions to the unitarity sum, and obtained from them exact results on quantities of physical interest. Improvements of the method before the creation of the Standard Model have been realized in [3, 4].

An important step forward was achieved by Bourrely, Machet and de Rafael [5], who noted that the field correlator used as input for obtaining bounds on the K_{l3} form factors can be evaluated reliably in the deep Euclidean region by perturbative QCD. By this, the modern approach clarified the issue of the number of subtractions required in the dispersion relation for the correlator, which was not always treated correctly in the studies performed before the advent of QCD.

© The Author(s), under exclusive licence to Springer Nature Switzerland AG 2019   97
I. Caprini, *Functional Analysis and Optimization Methods in Hadron Physics*,
SpringerBriefs in Physics, https://doi.org/10.1007/978-3-030-18948-8_5

We sketch below the main idea of the derivation of the Meimnn-Okubo bounds in the modern context. We consider a vector V_μ or an axial A_μ quark-transition current

$$J_\mu \equiv \bar{q}_2 \Gamma_\mu q_1, \qquad (5.1)$$

where q_i are spinors associated to heavy or light quarks, and let $\Pi_J^{\mu\nu}$ be the two-point momentum-space Green's function separated into manifestly spin-1 (Π_J^T) and spin-0 (Π_J^L) terms:

$$\Pi_{\mu\nu}^J(q) \equiv i \int d^4x \, e^{iqx} \langle 0 | T\{J_\mu(x) J_\nu^\dagger(0)\} | 0 \rangle$$

$$= \left(\frac{q_\mu q_\nu}{q^2} - g_{\mu\nu} \right) \Pi_J^T(q^2) + \frac{q_\mu q_\nu}{q^2} \Pi_J^L(q^2). \qquad (5.2)$$

The functions $\Pi_J^{T,L}$ satisfy dispersion relations with positive spectral functions, expressed by unitarity in terms of contributions from a complete set of hadronic states. The asymptotic behavior predicted by perturbative QCD implies that the dispersion relations require subtractions. The subtraction constants disappear by taking the derivatives of the amplitudes, for instance by defining the functions[1]

$$\chi_J^T(q^2) \equiv \frac{1}{2} \frac{\partial^2 \Pi_J^T}{\partial(q^2)^2} = \frac{1}{\pi} \int_0^\infty dt \, \frac{\mathrm{Im}\, \Pi_J^T(t)}{(t - q^2)^3},$$

$$\chi_J^L(q^2) \equiv \frac{\partial \Pi_J^L}{\partial q^2} = \frac{1}{\pi} \int_0^\infty dt \, \frac{\mathrm{Im}\, \Pi_J^L(t)}{(t - q^2)^2}. \qquad (5.3)$$

Perturbative QCD can be used to compute the functions $\chi_J(q^2)$ at values of q^2 far from the region where the current J^μ can produce manifestly nonperturbative effects like pairs of hadrons. When one of the quarks are charm or bottom quarks, a reasonable choice is $q^2 = 0$, while for the light quarks a nonzero spacelike value, like $q^2 = -2\,\mathrm{GeV}^2$, is suitable.

The spectral functions $\mathrm{Im}\, \Pi_J^{T,L}$ are evaluated by unitarity, inserting into the unitarity sum a complete set of states X that couple the current J to the vacuum:

$$\mathrm{Im}\, \Pi_{\mu\nu}^J(q^2) = \frac{1}{2} \sum_X (2\pi)^4 \delta^4(q - p_X) \langle 0 | J_\mu | X \rangle \langle X | J_\nu^\dagger | 0 \rangle. \qquad (5.4)$$

For most purposes, it is enough to keep in the sum over X the lightest meson pair in which one of meson (of mass m) contains the quark q_1 and the other (of mass m') contains the quark $\bar{q}_2$, and use the positivity of the higher-mass contributions. This choice gives a rigorous lower bound on the spectral functions, in terms of the form factors that parametrize the matrix elements of the current. For instance, for

---

[1] Other choices of invariant amplitudes which satisfy unsubtracted dispersion relations have been also investigated in the literature.

pseudoscalar initial and final mesons, the form factors associated to the vector weak current V_μ are defined through the matrix element

$$\langle M'(p')|V_\mu(0)|M(p)\rangle = \left(p_\mu + p'_\mu - q_\mu \frac{m - m'^2}{q^2} \right) f_+(t) + q_\mu \frac{m^2 - m'^2}{q^2} f_0(t),$$

$$(5.5)$$

where m and m' are the meson masses, $q = p - p'$, $t = q^2$ and $f_+(t)$ and $f_0(t)$ denote the vector and scalar form factor, respectively.

By combining the lower bounds on the spectral functions obtained from (5.4) with the dispersion relations (5.3), we obtain integral conditions expressed in terms of the modulus squared of the relevant form factors. For instance, the condition satisfied by the vector form factor $f_+(t)$ reads

$$\frac{1}{\pi \chi_J^T(q^2)} \int_{t_+}^{\infty} w_T(t) |f_+(t)|^2 \, dt \le 1,$$

$$(5.6)$$

where $w_T(t)$ is a product of phase-space factors entering the unitarity sum and the denominator of the first dispersion relation (5.3):

$$w_T(t) \sim \frac{(t - t_+)^{3/2}(t - t_-)^{3/2}}{t^3(t - q^2)^3}, \qquad t_\pm = (m \pm m')^2.$$

$$(5.7)$$

A similar inequality is obtained from the second dispersion relation (5.3) for the scalar form factor $f_0(t)$. We conclude that Meiman-Okubo method leads to model-independent constraints of the generic form (4.6). Using in addition the fact that the weak form factors $f_+(t)$ and $f_0(t)$ are analytic functions in the complex t plane cut for $t \ge t_+$ (this can be shown using the dispersion techniques in quantum field theory reviewed in Sect. 1.3), we can use the optimization problems presented in Sects. 4.1–4.5 for obtaining useful constraints on values of physical interest.

More generally, by including in the summation (5.4) also the contributions of states of higher invariant masses t_n, we obtain instead of (4.6) a relation of the form

$$\frac{1}{\pi} \sum_n \int_{t_n}^{\infty} w_n(t) |F_n(t)|^2 dt \le 1,$$

$$(5.8)$$

involving several form factors $F_n(t)$ which contribute above the specific thresholds t_n. Applications of this more general inequality will be considered below.

Since the unitarity relation (5.4) and the positivity of the spectral function play the major role in the above derivation, the Meiman-Okubo approach is also known as the formalism of "unitarity bounds".

5.2 Subthreshold Poles

A slight complication in the derivation of bounds starting from the L^2-norm condition
(5.6) is the fact that the form factors might have singularities on the real axis below
the unitarity threshold t_+. In some cases, the subthreshold singularities of the form
factors can be approximated by poles situated on the real axis. For instance, poles are
present in the form factors relevant for the weak semileptonic decays $B \to D^{(*)} \ell \bar{\nu}$
and $B \to \pi \ell \bar{\nu}$, while no poles are encountered in the form factors of the $K \to \pi \ell \bar{\nu}_\ell$
and $D \to \pi \ell \bar{\nu}_\ell$ decays. As mentioned above in Sect. 4.10, some authors argued that
the subthreshold poles, neglected in the bounds derived in [6] on the Isgur-Wise
function, would make the derivation of bounds impossible.

However, as shown in the subsequent works [7–9], the formalism can be applied
also in cases when poles are present, if their positions are known, even if the residues
are not known. The solution involves the Blaschke function which we used several
times in this book. Let us assume that a vector form factor $f_+(t)$ has a pole at $t_p < t_+$
and let $z_p = \tilde{z}(t_p, t_0)$ be the position of the pole in the z-plane, where $\tilde{z}(t, t_0)$ is the
conformal mapping (4.1). We consider the Blaschke factor

$$B(z; z_p) \equiv \frac{z - z_p}{1 - z_p^* z}, \tag{5.9}$$

which is analytic in $|z| \leq 1$, vanishes at $z = z_p$ and has modulus unity for z on the
unit circle:

$$|B(\zeta; z_p)| = 1, \qquad \zeta = e^{i\theta}. \tag{5.10}$$

We define then the new function

$$\tilde{f}_+(t) = B(\tilde{z}(t, t_0); z_p) f_+(t). \tag{5.11}$$

Since the pole present in $f_+(t)$ is cancelled by the zero of the Blaschke factor, the
function $\tilde{f}_+(t)$ is analytic in the t-plane cut for $t \geq t_+$. Moreover, from (5.10) it
follows that $\tilde{f}_+(t)$ has the same modulus on the cut as $f_+(t)$, therefore it satisfies the
same boundary condition, namely

$$\frac{1}{\pi \chi_J^T(q^2)} \int_{t_+}^{\infty} \frac{w_T(t) |\tilde{f}_+(t)|^2}{(t - q^2)^3} \, dt \leq 1. \tag{5.12}$$

Using the methods presented in Sect. 4.4, one can derive bounds for the analytic
function $\tilde{f}_+(t)$ and then, using (5.11), translate them in terms of the physical function
$f_+(t)$.

The presence of a pole leads to weaker bounds on the physical form factor at
points inside the analyticity domain, compared to the situation when the pole would
be absent. Indeed, for a single point the constraint similar to (4.65) takes now the
form

$$|f_+(t_1)| \leq \frac{1}{|B(z_1; z_p)\phi(z_1)|} \frac{1}{\sqrt{1 - z_1^2}}, \tag{5.13}$$

where $\phi(z)$ is the outer function related to the condition (5.12). Since $|B(z_1; z_p)| < 1$ for $|z_1| < 1$, the allowed range for $f_+(t_1)$ obtained from (5.13) is larger than the range given by the same formula without this factor. In particular, for points z_1 close to the pole position z_p, when $B(z_1; z_p)$ is very small, the constraint (5.13) becomes very weak.

On the other hand, if the residue r_p of the pole is known, using $f_+(t) \sim r_p/(t - t_p)$ for $t \sim t_p$ in (5.11), we obtain the value $\tilde{f}_+(t_p)$

$$\tilde{f}_+(t_p) = \frac{r_p}{1 - z_p^2} \left[\frac{d\tilde{z}(t, t_0)}{dt} \right]_{t=t_p}, \tag{5.14}$$

which can be used as an additional constraint, leading to improved bounds on the form factor at other values of t.

5.3 Semileptonic Form Factors

The Meiman-Okubo approach proved its utility especially for obtaining constraints on the form factors which describe the semileptonic decays of hadrons, in particular on the exclusive decays $M \to M'\ell\bar{\nu}_\ell$, where M and M' are pseudoscalar or vector mesons. Examples are the decays $K \to \pi\ell\bar{\nu}_\ell$, $D \to \pi\ell\bar{\nu}_\ell$, $B \to \pi\ell\bar{\nu}_\ell$ and $B \to D^{(*)}\ell\bar{\nu}_\ell$. The weak form factors involved in these decays are of central importance in strong interaction dynamics, providing information on the nature of the strong force and confinement. Phenomenologically, they are of crucial importance for the determination of standard model parameters such as the elements V_{us}, V_{cs}, V_{ub} and V_{cb} of the Cabibbo–Kobayashi–Maskawa (CKM) matrix.

As discussed above, the semileptonic form factors $f_+(t)$ and $f_0(t)$ are analytic functions in the complex t plane, with a unitarity cut along the real axis for $t \geq t_+ = (m + m')^2$. Angular momentum conservation imposes the behavior Im $f_+(t) \sim (t - t_+)^{3/2}$ near the threshold. The form factors satisfy in addition the Schwarz reflection condition, written generically as $F(t^*) = F^*(t)$, and are therefore real on the real t axis below t_+, in particular in the physical range of semileptonic decays, $m_\ell^2 \leq t \leq t_-$, where $t_- = (m - m')^2$. In this region, the functions $f_+(t)$ and $f_0(t)$ determine the decay rate by

$$\frac{d\Gamma}{dt} = \frac{G_F^2|V_{ij}|^2}{192\pi^3 m^3} \frac{k}{t^{\frac{5}{2}}} (t - m_\ell^2)^2 \left[4k^2 t(2t + m_\ell^2)|f_+(t)|^2 + 3m_\ell^2|f_0(t)|^2 \right], \tag{5.15}$$

where $k = 1/2\sqrt{(t_+ - t)(t_- - t)/t}$, G_F is the Fermi constant and V_{ij} is the relevant element of the CKM matrix.

Using the techniques discussed in Sects. 4.4–4.5, completed when necessary with the recipe of treating the subthreshold poles presented in Sect. 5.2, one can derive constraints on the semileptonic form factors defined above. The problem has been investigated in a large number of works [2–27]. The aim of most recent studies is to obtain nontrivial constraints on the analytic parametrizations of the form factors in the physical region and to find estimates of the truncation errors. The continuous progress in the calculations by perturbative QCD of the correlators $\chi(Q^2)$ (known at present in some cases to order α_s^4), and additional theoretical input from χPT, HQET and lattice QCD, led to a gradual increase in the precision of the model-independent form-factor parametrizations.

Several types of parametrizations have been used for the semileptonic form factors in the context of unitarity bounds. We shall briefy discuss here two of them (other choices will be presented in Sects. 5.3.3 and 5.3.4).

Let $F(t)$ be a generic form factor. A parametrization often adopted in the literature is based on the Taylor expansion at $t = 0$

$$F(t) = \sum_{k \geq 0} c_k t^k, \qquad (5.16)$$

which converges in the disc $|t| < t_+$ limited by the first unitarity branch point $t_+ = (m + m')^2$. In the semileptonic range $m_l^2 \leq t \leq t_-$, where $t_- = (m - m')^2$, the asymptotic convergence rate scales as t_-/t_+.

An alternative parametrization used in applications is

$$F(t) = \frac{1}{\phi(z) B(z; z_p)} \sum_{k \geq 0} g_k z^k, \qquad (5.17)$$

where $z = \tilde{z}(t, t_0)$ is the conformal variable (4.1), $\phi(z)$ is the outer function defined in Sect. 4.2 and $B(z; z_p)$ is the Blaschke factor accounting for the subthreshold poles, as discussed in Sect. 5.2 (this factor is absent if the form factor has no poles). The expansion (5.17) converges in the whole disk $|z| < 1$, i.e. in the whole t plane up to the unitarity cut situated on the real axis for $t \geq t_+$.

The extremal problems discussed in Sect. 4.4 allow one to derive constraints on the first real coefficients c_k appearing in (5.16) or on the coefficients g_k appearing in (5.17), depending on the input available for the form factor of interest. These constraints are useful for improving the fits of the experimental decay rates.

One might ask how accurate are the representations of the form factors in the physical region by a finite number of terms. Without additional information, it is in general impossible to estimate the magnitude of the higher terms in an expansion, even if it is convergent. However, the model-independent condition of the generic form (4.6), derived by Meiman-Okubo technique, is useful for constraining not only the first coefficients of the expansions, but also the truncation error.

Keeping in the expansion (5.16) terms with $k \leq K - 1$, the truncation error can be taken as the magnitude of the first neglected term:

$$\delta F(t)_{\text{trunc}} \sim |c_K t^K|.\tag{5.18}$$

Using the methods presented in Sects. 4.4–4.5, one can find upper and lower bounds on the coefficient c_K for fixed values of the lower-order coefficients $c_0, \ldots, c_{K-1}$, which gives an estimate of the truncation error (5.18).

For the expansion (5.17), keeping again terms with $k \le K - 1$, one can define a conservative truncation error

$$\delta F(t)_{\text{trunc}} = \left| \frac{1}{\phi(z)B(z; z_p)} \sum_{k=K}^{\infty} g_k z^k \right|,\tag{5.19}$$

which accounts for all the neglected terms in the expansion. By applying Cauchy-Schwarz inequality and taking into account the condition

$$\sum_{k=K}^{\infty} g_k^2 \le 1 - \sum_{k=0}^{K-1} g_k^2,\tag{5.20}$$

which follows from the L^2 norm condition (4.52), we obtain the exact upper bound

$$\delta F(t)_{\text{trunc}} \le \frac{1}{|\phi(z)B(z; z_p)|} \left(1 - \sum_{k=0}^{K-1} g_k^2 \right) \frac{|z^K|}{\sqrt{1 - z^2}}, \quad z = \tilde{z}(t, t_0),\tag{5.21}$$

written in terms of the values of the first K coefficients obtained from data.

Alternatively, the error attached to (5.17) can be taken as

$$\delta F(t)_{\text{trunc}} \le \frac{1}{|\phi(z)B(z; z_p)|} |g_K z^K|, \quad z = \tilde{z}(t, t_0),\tag{5.22}$$

which is estimated by finding upper and lower bounds on the coefficient g_K for fixed values of the lower-order coefficients $g_0, \ldots, g_{K-1}$ and the other constraints satisfied by the form factor.

The errors defined by (5.21) and (5.22) are minimized on the whole semileptonic region by mapping this region onto a narrow range around $z = 0$ in the z plane. It is easy to see that this is achieved by choosing the parameter t_0 in the conformal mapping (4.1) such as to map the region $(0, t_-)$ onto an interval $(-z_{\max}, z_{\max})$ symmetric around the origin in the z plane. From the condition $\tilde{z}(0, t_0) = -\tilde{z}(t_-, t_0)$ one obtains the optimal value

$$t_0 = (m + m')^2 (\sqrt{m} - \sqrt{m'})^2.\tag{5.23}$$

In practice, it is reasonable to increase the number K of terms in the expansion until the bound on the truncation error obtained from (5.21) or (5.22) becomes smaller than the experimental uncertainty.

In the next subsections we shall briefly discuss the form factors relevant for the decays $K \to \pi \ell \bar{\nu}_\ell$, $D \to \pi \ell \bar{\nu}_\ell$, $B \to \pi \ell \bar{\nu}_\ell$ and $B \to D^{(*)} \ell \bar{\nu}_\ell$. In each case we shall point out the questions of interest, the specific extremal problem used for solving them and the main results. Details and other applications can be found in the articles quoted in each case.

5.3.1 $K\pi$ Form Factors

The first applications of the Meiman-Okubo formalism [2–4], in the early 1970s, have been performed on the K_{l3} form factors, which describe the semileptonic decays $K \to \pi \ell \bar{\nu}_\ell$. The approach was revisited after the advent of QCD in [5], and more recent studies have been done in [17, 18, 22–24]. While in the early works the formalism was used mainly for calculating upper and lower bounds on the decay rate (5.15), the aim of the recent studies is to improve the precision of the parametrizations of the form factors used for data analysis in the semileptonic region.

The parametrizations used traditionally in this case are based on the Taylor expansion (5.16), written as

$$f_i(t) = f_i(0) \left[1 + \lambda_i' \frac{t}{m_\pi^2} + \lambda_i'' \frac{t^2}{2m_\pi^4} + \cdots \right], \qquad i = +, 0, \qquad (5.24)$$

in terms of dimensionless parameters (slope λ_i' and curvature λ_i''). The expansions converge in the disk $|t| < t_+$ limited by the branch point $t_+ = (m_K + m_\pi)^2 = 0.401\,\text{GeV}^2$. In the semileptonic range $m_l^2 \le t \le t_-$, where $t_+ = (m_K - m_\pi)^2 = 0.125\,\text{GeV}^2$, the convergence is expected to be rather good, with the asymptotic convergence rate $t_-/t_+ = 0.31$. At the present experimental accuracy, a theoretical correlation between the coefficients λ_i' and λ_i'' proves to be very helpful.

A model-independent correlation can be derived by using the Meiman-Okubo constraint (5.6) for the vector form factor $f_+(t)$ (and a similar one for $f_0(t)$) and the phases, $\arg f_i(t + i\varepsilon)$, known in the elastic region by Fermi-Watson theorem from the $I = 1/2$ S and P-wave phase shifts of elastic $K\pi$ scattering. The first inelastic threshold, due to $K\eta/K^*\pi$ channels for the scalar/vector form factor, can be taken in both cases at $t_{in} \approx 1$ GeV^2. In addition, low-energy theorems predict the values $f_+(0) = f_0(0)$ and, for the scalar form factor, also the value at the Callan-Treiman point $\Delta_{K\pi} = m_K^2 - m_\pi^2$. The explicit form of the low-energy constraints, the perturbative QCD correlators and the phenomenological input can be found in [23].

Using the extremal problem solved in Sect. 4.4.3, one can derive a model independent relation between the slope and the curvature defined in (5.24). As discussed [17, 23], the solution (4.62) of the relevant extremal problem can be expressed as a quadratic convex inequality in terms of λ_i' and λ_i'', so that the allowed domain in the plane of these parameters is the interior of an ellipse. The constraints can be further improved by noting that there is also information on the modulus of the form factors

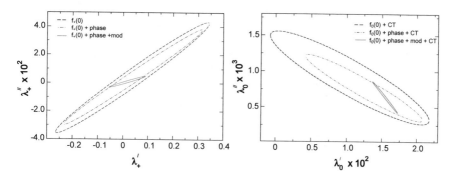

Fig. 5.1 Allowed domains for the slope and curvature of the vector (left) and scalar (right) $K\pi$ form factors

below t_{in}, from the rate of the $\tau \to K\pi\nu$ decay. Therefore, one can use the extremal problem formulated in Sect. 4.4.4, which gives a tighter correlation between the slope and the curvature.

For illustration we show in Fig. 5.1 the allowed domains for the parameters λ_i' and λ_i'', derived in [23], delimited by the corresponding ellipses. The largest domains are obtained using the Meiman-Okubo bound of the generic form (4.6), the normalization $f_+(0) = f_0(0) = 0.962$, and (in the scalar case) the CT condition $f_0(\Delta_{K\pi}) = 1.193$. The reduction of the domains by using also the phase and phase and modulus information up to $t_{in} = 1$ GeV^2 is illustrated by the smaller ellipses. The figures show the increasing constraining power of the additional information used as input. A strong correlation between the slope and the curvature is obtained in all cases.

5.3.2 $D\pi$ Form Factors

The form factors describing the decay $D \to \pi\ell\bar{\nu}$ have been studied with the Okubo-Meiman technique in [10, 25], and more recently in [26, 27]. In this case, the unitarity threshold and the end of the semileptonic region are $t_+ = 4.02$ GeV^2 and $t_- = 2.98$ GeV^2, respectively (using the masses of the neutral D and charged π mesons). The large value of t_+ justifies the choice $q^2 = 0$ in the dispersion relations (5.3), as discussed in Sect. 5.1.

As in the case of $K\pi$ form factors, the low-energy theorems predict the values $f_+(0) = f_0(0)$ and, for the scalar form factor, also the value at the Callan-Treiman point $\Delta_{D\pi} = m_D^2 - m_\pi^2$. On the other hand, Watson-Fermi theorem is not useful in the case of the $D\pi$ form factors, since $D\pi$ elastic scattering is poorly known. However, from the shape of the lowest scalar resonance $D_0^*(2403)$ one can infer the phase of the scalar form factor below the inelastic threshold $t_{in} = (2.6\,\text{GeV})^2$, given by the opening of $D\eta$ channel (for the vector form factor, the resonance $D^*(2010)$ is very close to the threshold and does not allow a reasonable extraction of the phase). Details on the input are given in [25].

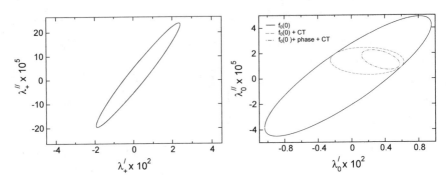

Fig. 5.2 Allowed domains for the slope and curvature of the vector (left) and scalar (right) $D\pi$ form factors

In Fig. 5.2, we show for illustration the allowed domains obtained in [25] for the slope and curvature of the $D\pi$ form factors, defined as in (5.24). The domains are obtained using the Meiman-Okubo bound of the generic form (4.6) and the normalization $f_+(0) = f_0(0) = 0.65$. In the scalar case the CT condition $f_0(\Delta_{D\pi}) = 1.58$ and the phase below $t_{in} = (2.6\,\text{GeV})^2$, obtained from the shape of the D_0^* resonance, are also implemented, leading to smaller allowed domains.

Since the ratio $t_-/t_+ = 0.74$ is rather large, the convergence of the expansion (5.24) for the $D\pi$ form factors is expected to be poor at the end of the semileptonic region. The alternative expansion (5.17) (with the Blaschke factor set to 1, since the $D\pi$ form factors have no subthreshold poles) has a better convergence. In this case, the Meiman-Okubo formalism leads to constraints on the coefficients g_k.

As an application, we indicate below the improvement of the constraints on these coefficients brought in the scalar case by the knowledge of the phase below t_{in}. We use in this case the solution (4.99) of the problem solved in Sect. 4.4.3. By applying this constraint to the scalar $D\pi$ form factor for $K = 5$, leads to an allowed domain for the coefficients g_k, $k \leq 4$, writen as [27]:

$$1.78\, g_0^2 + 1.28\, g_1^2 + 1.13\, g_2^2 + 1.84\, g_3^2 + 2.33\, g_4^2 + 0.79\, g_0 g_1$$
$$- 0.49\, g_0 g_2 - 1.61\, g_0 g_3 - 1.96\, g_0 g_4 - 0.13\, g_1 g_2 - 0.84\, g_1 g_3$$
$$- 1.09\, g_1 g_4 + 0.49\, g_2 g_3 + 0.54\, g_2 g_4 + 2.09\, g_3 g_4 \leq 1. \qquad (5.25)$$

From general arguments, one expects this constraint to be stronger than the general condition (4.52) obtained without input on the phase, which implies

$$\sum_{k=0}^{4} g_k^2 \leq 1. \qquad (5.26)$$

In order to assess numerically the improvement, we recall that in a typical application to semileptonic processes, the lowest coefficients g_k are determined from fits of the data, and the aim is to set a bound on the next coefficient g_K, which enters the truncation error (5.22). It turns out that, in practical fits, the optimal values of the parameters are usually small, far from saturating the upper bound (5.26). To simulate such a situation, we take, for instance, the input values $g_0 = 0.10$, $g_1 = 0.08$, $g_2 = 0.07$ and $g_3 = 0.05$, for which the left hand side of (5.26) is 0.024. With this input, we obtain the constraint $|g_4| \leq 0.99$ from the standard inequality (5.26), and the smaller range $-0.62 \leq g_4 \leq 0.68$ from the improved constraint (5.25). Using the above limits on g_4 and the values $z_- = 0.325$ and $\phi(z_-) = 0.176$ in the present case, we obtain from (5.22) the uncertainties $\delta F(t_-) \approx 0.063$ using the standard constraint (5.26) and $\delta F(t_-) \approx 0.043$ using the improved constraint (5.25), which amounts to an improvement by about 30%. Similar results are obtained for other input values for the lowest coefficients. We note that in the $D\pi$ case a further improvement of the truncation error is obtained by using the symmetric mapping of the semileptonic region around the origin in the z plane, which is achieved using in (4.1) the optimal value $t_0 = 1.97\,\mathrm{GeV}^2$ obtained from (5.23).

5.3.3 $B\pi$ Form Factors

The first applications of the Meiman-Okubo approach to the form factors that describe the decay $B \to \pi \ell \bar{\nu}$ have been performed in [10, 13, 14]. More recently the $B\pi$ form factors have been investigated in [19–21]. We shall discuss here the application of the formalism for improving the parametrization of the $B\pi$ vector form factor in the semileptonic region, of interest for the extraction of the element V_{ub} of the CKM matrix. As in the case of other form factors involving a heavy meson, it is reasonable to take $q^2 = 0$ in the condition (5.6).

Using the masses of the neutral B and charged π mesons, the unitarity threshold and the end of the semileptonic region are $t_+ = 29.37\,\mathrm{GeV}^2$ and $t_- = 26.42\,\mathrm{GeV}^2$, respectively. The large value of the ratio $t_-/t_+ = 0.9$ indicates that the Taylor expansion (5.16) is poorly convergent near the right end of the semileptonic region. Recalling that the vector form factor $f_+(q^2)$ has a subthreshold pole due to the B^* state of mass $m_{B^*} = 5.325\,\mathrm{GeV}$, one can use the expansion in powers of the z variable of the form (5.17), including the corresponding Blaschke factor. This parametrization was investigated in [19, 20].

However, as noticed in [19] and discussed in detail in [21], when the expansion (5.17) is truncated at a finite order, the form factor has a bad behavior both at the threshold $t = t_+$ and at large $|t|$. This follows from the expression (4.24) of the outer function. Using the expression (4.24), one can see that in the present case the exponents in (4.24) are $k = 3/2$, $l = 3/2$, $m = -3/2$ and $n = -3$, which implies that

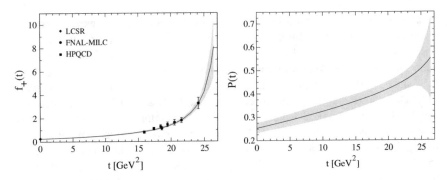

Fig. 5.3 Left: the vector $B\pi$ form factor $f_+(t)$. Right: the numerator $P(t) = f_+(t)(1 - t/m_{B^*}^2)$ of (5.28)

$$\frac{1}{\phi(z)} \sim \frac{1}{(1 - z)^2}, \qquad \text{for } z \sim 1,$$
$$\frac{1}{\phi(z)} \sim \frac{1}{(1 + z)^{1/2}}, \qquad \text{for } z \sim -1. \tag{5.27}$$

From the expression (4.1) of z, it follows that $1/\phi$ behaves as $1/(t_+ - t)$ at the $B\pi$ production threshold t_+, and increases like $t^{1/4}$ at large $|t|$. This behavior is transferred to the form factor $f_+(t)$, as follows from (5.17) when the expansion in powers of z is truncated. The unphysical singularity at threshold may distort the behavior near the upper end of the semileptonic region, and the asymptotic behavior is in contradiction with perturbative QCD scaling which requires a decrease like $1/t$.

It should be noted that in the calculation of bounds one uses the full expansion in (5.17), with an infinite number of terms. Then the series cancels the zeros of the function $\phi(z)$ at $z = \pm 1$, restoring the required properties of $f_+(t)$. Actually, it can be shown that by imposing the condition that the series vanishes at threshold or at infinity does not change the unitarity bounds.

In order to avoid these shortcomings, we consider the following parametrization

$$f_+(t) = \frac{1}{1 - t/m_{B^*}^2} \sum_{k=0}^{K-1} b_k(t_0) z^k, \quad z = \tilde{z}(t, t_0), \tag{5.28}$$

which was proposed in [21]. It can be easily checked that the expression (5.28) has the correct analytic structure in the complex t plane and the proper scaling, $f_+(t) \sim 1/t$ at large $|t|$. The threshold condition $\mathrm{Im}\, f_+(t) \sim (t - t_+)^{3/2}$ near t_+, mentioned above, is imposed by requiring

$$\left[\frac{df_+}{dz}\right]_{z=1} = 0, \tag{5.29}$$

and can be expressed as a linear relation between the coefficients b_k. These coefficients satisfy also the unitarity constraint

$$\sum_{j,k=0}^{K} B_{jk}(t_0)b_j(t_0)b_k(t_0) \leq 1, \tag{5.30}$$

which follows from the standard condition (4.52) on the coefficients g_k of (5.17). The numbers $B_{jk}(t_0)$ have been calculated in [21].

The parametrization (5.28) with the additional constraints (5.29) and (5.30) was used in [21] for fitting the experimental information on the $B \to \pi \ell \bar{\nu}$ decay and the theoretical values calculated from lattice QCD. The constraint (5.30) has been used also to estimate the truncation error, defined as

$$\delta f_+(t)_{\text{trunc}} = \frac{b_K^{max} |z^K|}{1 - t/m_{B^*}^2}, \tag{5.31}$$

where b_K^{max} is the maximum value calculated from the inequality (5.30), for fixed values of the lower-order coefficients determined from the fit. For minimizing this error, the parameter t_0 defining the conformal mapping was set at the optimal value $t_0 = 20.06$ GeV^2 from (5.23). Moreover, the number of terms in the expansion was increased until the truncation error became smaller than the statistical one. It turns out that for $K = 4$ this is valid along the whole semileptonic region. The optimal values of the parameters b_k and of the error are given in [21].

The form factor calculated in Ref. [21] using the optimal parameters is shown in Fig. 5.3. The error band is given by the statistical uncertainties, since the systematic errors are negligible in the semileptonic domain. The theoretical LCSR result from [28] and the lattice results from [29, 30] are also shown. The right panel shows the z polynomial in the numerator of (5.28) for the optimal parameters and their statistical errors. The unitarity constraint plays a nontrivial role, being responsible for the asymmetric errors, especially near the right end of the semileptonic range. As discussed in [21] and in many papers afterwards, the parametrization (5.28) of the $B\pi$ vector form factor and the error analysis presented above allowed a precise prediction for the element $|V_{ub}|$ of the CKM matrix from exclusive $B \to \pi$ decays.

5.3.4 $BD^{(*)}$ Form Factors

The form factors which describe the decays $B \to D\ell\bar{\nu}$ and $B \to D^*\ell\bar{\nu}$ are crucial for the extraction of the element V_{cb} of the CKM matrix. They have been investigated in the frame of Meiman-Okubo formalism in [11, 12, 15, 16]. Recently, the interest in this approach increased in connection with the debates on the role of heavy-quark symmetry in the extraction of V_{cb} from exclusive decays [31–36]. The problem is of interest also for understanding the experimental ratio of the rates of $B \to D\ell\bar{\nu}$ decays for light and τ leptons, the so-called $R(D)$ anomaly.[2]

For $B^{(*)} \to D^{(*)}$ form factors, the formalism presented in Sect. 5.1 was applied for both vector $V_\mu = \bar{c}\gamma_\mu b$ and axial $A_\mu = \bar{c}\gamma_\mu \gamma_5 b$ currents. Moreover, in each case one can keep the states $|BD\rangle$, $|BD^*\rangle$, $|B^*D\rangle$, and $|B^*D^*\rangle$ in the unitarity relation (5.4) for the spectral functions of the correlators. This leads to four relations of the form (5.8), for the spin–parity channels $J^P = 0^+$, 0^-, 1^- and 1^+, each involving several form factors. These form factors behave as independent functions on the unitarity cut, but are related near the point $t = t_-$ of zero recoil by heavy-quark symmetry. This additional correlation considerably increases the constraining power of the Meiman-Okubo formalism.

It is convenient to work with the dimensionless kinematical variable $w = v \cdot v'$, where v and v' are the meson velocities, which is related to the momentum transfer squared t by:

$$t = m_{B^{(*)}}^2 + m_{D^{(*)}}^2 - 2m_{B^{(*)}} m_{D^{(*)}} w. \tag{5.32}$$

In terms of w, the various $B^{(*)} D^{(*)}$ thresholds[3] all occur at the same value $w = -1$. This enables a unified treatment of all form factors by a single conformal mapping

$$\tilde{z}(w, a) = \frac{\sqrt{w+1} - \sqrt{2}\, a}{\sqrt{w+1} + \sqrt{2}\, a}; \quad a > 0, \tag{5.33}$$

which maps the branch point $w = -1$ onto $z = -1$. We note that for $a = 1$ the zero recoil point $w = 1$ is mapped onto the origin $z = 0$, while for the optimal choice $a_{opt} = 1.509$ the semileptonic region is mapped onto the narrow symmetric range $(-z_{max}, z_{max})$ around the origin, with $z_{max} = 0.032$. Then the Meiman-Okubo-type relation (5.8) leads to four unitarity inequalities of the form

$$\frac{1}{2\pi} \int\limits_0^{2\pi} d\theta \sum_j |\phi_j(e^{i\theta})\, F_j(w)|^2 \le 1, \tag{5.34}$$

where $\phi_j(z)$ are outer functions and $F_j(w)$ are form factors contributing in each channel of definite spin and parity J^P. From these inequalities, one can derive constraints of the form

$$\sum_j \sum_{k=0}^\infty (g_k^{(j)})^2 \le 1, \tag{5.35}$$

---

[2] As shown in [37], the world average of $R(D)$ and $R(D^*)$ measured by *BABAR*, Belle and LHCb is in tension with the SM expectation at the 4σ level.

[3] For the form factors of interest, the thresholds occur at the minimum values of $\sqrt{t}$ at which the relevant $B^{(*)} D^{(*)}$ pairs can be produced, i.e. at $(m_B + m_D) \approx 7.15\,\text{GeV}$, $(m_B + m_{D^*}) \approx 7.29\,\text{GeV}$, $(m_{B^*} + m_D) \approx 7.19\,\text{GeV}$, or $(m_{B^*} + m_{D^*}) \approx 7.33\,\text{GeV}$.

where the coefficients $g_k^{(j)}$ are defined by the expansions of the type (5.17)

$$F_j(w) = \frac{1}{\phi_j(z)B_j(z, z_p)} \sum_{k=0}^{\infty} g_k^{(j)} z^k. \tag{5.36}$$

Note that subthreshold poles are present in some of the form factors, requiring their treatment by the Blaschke factor in the denominator of (5.36).

The strength of the inequality (5.35) comes from heavy-quark symmetry, which relates the different form factors near the point $w = 1$ of zero recoil. In order to incorporate the corrections to the heavy quark limit, it is convenient to express the form factors as products of a reference form factor and a calculable function. In [16] the reference form factor was taken to be the function $V_1(w)$ defined as

$$V_1(w) = h_+(w) - \frac{1-r}{1+r} h_-(w), \qquad r = \frac{m_D}{m_B}, \tag{5.37}$$

in terms of the $h_\pm(w)$ form factors parametrizing the particular matrix element

$$\langle D(v')|V_\mu|\bar{B}(v)\rangle = h_+(w)(v+v')_\mu + h_-(w)(v-v')_\mu. \tag{5.38}$$

Choosing V_1 as reference form factor, we define the ratios

$$R_j(w) \equiv \frac{F_j(w)}{V_1(w)} \equiv A_j[1 + B_j(w - w_0) + C_j(w - w_0)^2 + D_j(w - w_0)^3 + \cdots], \tag{5.39}$$

where the parameters $A_j, \ldots, D_j$ include short distance and $1/m_Q$ corrections [16].

A convenient parametrization for $V_1(w)$ is the Taylor expansion

$$V_1(w) = V_1(w_0)\left[1 - \rho_1^2(w - w_0) + c_1(w - w_0)^2 + d_1(w - w_0)^3 + \cdots\right], \tag{5.40}$$

where w_0 is such that $\tilde{z}(w_0, a) = 0$. One can choose $a = 1$, when the zero-recoil point $w = 1$ is mapped to $z = 0$, since the expansion (5.40) has a good convergence in the physical range, ensured by the small values of the ratios t_-/t_+ and t_-/t_p, where t_p is the position of the nearest subthreshold pole (for instance, in the BD case, $t_-/t_+ = 0.23$ and $t_-/m_{B^*}^2 = 0.29$).

However, a much better convergence is obtained by expanding in powers of the conformal variable $z = \tilde{z}(w, a_{opt})$ defined above, when the ratio of convergence $r_{conv} = z_{max}/z_p$ is much smaller, $r_{conv} \approx 0.07$. A simple algebraic calculation relates the coefficients of the expansions in different variables, so that we have

$$\frac{V_1(w)}{V_1(w_0)} = 1 - 8 a_{opt}^2 \rho_{1*}^2 z + 16 a_{opt}^2 (4a_{opt}^2 c_{1*} - \rho_{1*}^2) z^2 + \cdots, \tag{5.41}$$

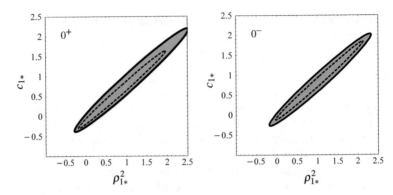

Fig. 5.4 Bounds on the slope and curvature of the reference form factor $V_1(w)$ at $w_0 = 1.276$. The solid (dashed) ellipses show the results obtained with (without) including theoretical uncertainties in the heavy-quark symmetry relations

where the parameters ρ_{1*}^2 and c_{1*} are the slope and curvature in (5.40) for $w_0 = 1.276$, which results from $\tilde{z}(w_0, a_{opt}) = 0$.

Using (5.39) and the unitarity constraints (5.35), one can obtain an allowed domain for the parameters ρ_{1*}^2 and c_{1*} for each J^P channel. In Fig. 5.4 we indicate for illustration the domains obtained in Ref. [16] from the scalar and pseudoscalar channels, which provide the strongest constraints. As discussed in detail in [16], the unitarity constraints (5.35) give also a stringent bound on the truncation error of the expansion (5.41).

As we mentioned above, the interest in the parametrizations of the $BD^{(*)}$ form factors increased again recently. In this context, the role of heavy-quark symmetry in the Meiman-Okubo formalism was reevaluated in detail in [31–33, 35, 36]. The problem is still under active investigation.

5.4 Pion Electromagnetic Form Factor

The electromagnetic form factor of the pion, $F_\pi^V(t)$ is a fundamental observable of the strong interactions and a sensitive probe of the composite nature of the pion. We defined this quantity in (1.45) by the matrix element of the electromagnetic current between charges pion states.[4] In Sect. 1.3.1, the derivation of analyticity properties from axiomatic field theory was reviewed using for illustration this form factor. We also mentioned the pion vector form factor in Sect. 2.2 in connection with the non-standard input in dispersion relations and in the extremal problem formulated in Sect. 2.4.1.

---

[4]A similar quantity, not observable but of theoretical interest for χPT, is the pion scalar form factor, defined from the matrix element between pion states of a suitable scalar operator. The pion scalar form factor was investigated in the Meiman-Okubo approach in [38, 39].

Due to the extensive experimental and theoretical information, the pion vector form factor is, compared with other hadronic quantities, a well-known function. However, the precision does not reach the same level for all timelike and spacelike momenta. On the timelike axis, in the elastic region $t < t_{in}$, the phase is well known, being equal by Fermi-Watson theorem to the P-wave phase shift $\delta_1^1(t)$ of the $\pi\pi$ scattering, which has been calculated with high precision using ChPT and Roy equations (see Refs. [11–13] quoted in Chap. 2). However, the modulus, extracted from the cross section of the $e^+e^- \to \pi^+\pi^-$ process [40–47], is affected at low energies by larger errors, due to difficulties of the experimental measurements. One may hope to improve the knowledge in these regions by the analytic continuation from regions where the precision is better.

As we saw in Sect. 1.5, the standard dispersion theory for form factors is based on the Omnès representation (1.89). It requires the knowledge of the phase along the whole cut for calculating the function $\Omega(t)$ and the positions of the zeros in the complex plane, entering the polynomial $P(t)$. This information is only partially available: above the inelastic threshold t_{in} the phase of the form factor is not known. As concerns the zeros, in [48] it is argued that the pion charge distribution should be similar to that of the electron in the ground state of the hydrogen atom, being proportional to the square of the wave function. Therefore, the form factor is expected to be positive in the spacelike region. But the presence of zeros in the complex plane can not be excluded. One can avoid model-dependent assumptions in the evaluation of the Omnès representation (1.89) by using as input alternative information available from different sources.

The first application of the Meiman-Okubo formalism to the pion vector form factor was performed in [49]. However, the results (for instance, for the slope and curvature in the Taylor expansion (2.1)) turned out to be rather weak. Actually, for this form factor the experimental and theoretical information supersedes to a large extent the model-independent bound of the type (5.6) which is obtained by Meiman-Okubo method. From measurements of $|F_\pi^V(t)|$ for $t > t_{in}$ by BABAR Collaboration [42] and the asymptotic scaling $|F_\pi^V(t)| \sim 1/t$ predicted by perturbative QCD [50, 51], one can wite down an inequality of the form (2.6), for a suitably normalized weight $w(t)$.

Therefore, for the pion vector form factor we consider the extremal problem formulated already in Sect. 2.4.1, defined by Eqs. (2.21)–(2.23). This approach was applied in [52–57] for deriving bounds on the pion form factor in kinematical regions where it is less known. As remarked in these works, the constraining power of the formalism is increased by using as input also the values of the form factor at spacelike energies, measured by the recent experiments [58, 59], and the most precise values of the modulus at timelike energies, measured from the e^+e^- annihilation into two pions in Refs. [40–47]. Thus, we add to the constraints (2.21) defining the admissible class $\mathscr{F}$ the conditions

$$F(t_s) = F_s \pm \varepsilon_s, \qquad t_s < 0, \tag{5.42}$$

and

$$|F(t_t)| = F_t \pm \varepsilon_t, \qquad t_+ < t_t < t_{in}, \tag{5.43}$$

for a given spacelike point t_s and a given timelike point t_t.

The solution of the problem formulated in (2.21)–(2.23) and supplemented by the conditions (5.42)–(5.43) was given in Sect. 4.4.4. As explained there, the mixed phase and modulus problem was written finally in the form (4.107), i.e. it was reduced to an interpolation problem in L^2 norm, solved in Sect. 4.4. The solution of this problem, written as the inequality (4.62), describes finally the allowed domain of the values of the form factor and its derivatives at some interior points. From this inequality, one can derive upper and lower bounds on a quantity of interest, in terms of the remaining values entering the determinant (4.62). Consider, for instance, the particular inequality

$$\det \begin{bmatrix} 1 - g(0)^2 - g'(0)^2 & \bar{g}(z_a) & \bar{g}(z_b) \\ \bar{g}(z_a) & \frac{z_a^4}{1-z_a^2} & \frac{(z_a z_b)^2}{1-z_a z_b} \\ \bar{g}(z_b) & \frac{(z_a z_b)^2}{1-z_a z_b} & \frac{(z_b)^4}{1-z_b^2} \end{bmatrix} \geq 0, \tag{5.44}$$

where $z_a = \tilde{z}(t_a, 0)$, $z_b = \tilde{z}(t_b, 0)$, t_a and t_b are the spacelike and timelike points used as input, the function $g(z)$ is defined in (4.106) and $\bar{g}(z) = g(z) - g(0) - g'(0)z$. We note that the derivative $g'(0)$ is expressed in terms of the first derivative of the form factor at $t = 0$, which in turn is related to the charge radius by

$$\langle r^2 \rangle_\pi^V = 6 \left[d F_\pi^V(t)/dt \right]_{t=0}, \tag{5.45}$$

as follows from the Taylor expansion (2.1). Thus, we can derive from (5.44) upper and lower bounds on $\langle r_\pi^2 \rangle$ in terms of known quantities, by simply solving a quadratic inequality. Bounds on the form factor at other points can be found in a similar way from the determinant (4.62) with suitable entries.

A nontrivial complication is, however, the fact that the experimental values (5.42) and (5.43) are affected by statistical errors. This requires to devise a procedure of properly merging the formalism of exact bounds with the statistical distribution of some of the input values. If a parametrization of the form factor were used, the output errors on the parameters and their correlation matrix would be obtained by standard χ^2 minimization. But the formalism applied here uses no parametrizations: one must predict in a model-independent framework a central value and an uncertainty with a definite confidence level for the result.

The problem can be solved by generating a large sample of pseudo-data, achieved by randomly sampling each of the input quantities with specific distributions: the phase of the form factor, which is obtained from a theoretical calculation (solution of Roy equations for $\pi\pi$ scattering partial-wave amplitudes), is assumed to be uniformly distributed, while for each spacelike (5.42) and timelike (5.43) input it is reasonable to adopt a Gaussian distribution with the measured central value as mean and the quoted error as standard deviation.

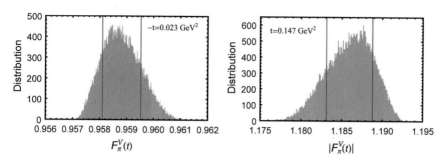

Fig. 5.5 Statistical distributions of the output values of the form factor at one spacelike and one timelike point, for a specific experimental input. The vertical lines indicate the 68.3% confidence limit (CL) intervals

For each point from the input statistical sample, if the values are compatible[5] the inequality (5.44) (or other particular case of (4.62)), gives upper and lower bounds on the quantity of interest, for instance the value of the form factor $F_\pi^V(t)$ at some point t, or the charge radius (5.45). Since all the values between the extreme points are equally allowed, uniformly-distributed values of the quantity of interest are then generated in between the bounds. For convenience, a minimal separation between the generated points must be set, and for allowed intervals smaller than this limit no intermediate points are created. In this way, a large sample of values for the output quantity is generated (for numerical details of the procedure see [55–57]). The distributions of the calculated values prove to be stable against the variation of the size of the random sample and the minimal separation between the generated points and have a Gaussian shape, allowing the extraction of the mean value and the standard deviation for the quantity of interest.

For illustration we present in Fig. 5.5 the statistical distributions of the values of the form factor at one spacelike point $t < 0$ and one timelike point $t > 0$, obtained in [57], and in Fig. 5.6 the statistical distribution of the charge radius squared $\langle r_\pi^2 \rangle$, obtained in [56] with a specific input (one value on the spacelike axis and one value on the timelike axis).

The final predictions are obtained by averaging the values given by different input data from a suitable timelike region, where the different experiments [40–47] are in general agreement. In [55–57], this region was chosen as 0.65–0.7 GeV. A method based on the determination of correlations from data [60] was applied for a conservative calculation of the uncertainties.

In Fig. 5.7 we show the results obtained in [57] for the pion vector form factor in two kinematical regions where the previous knowledge was rather poor. In the left panel, the form factor at small spacelike t is compared with the experimental results of the NA7 Collaboration [61] and the lattice calculations of the ETM Collaboration [62]. In the right panel the modulus of the form factor on the cut below

---

[5]As discussed below (4.62), the minors of the determinant should be also nonnegative. Some of these conditions involve only input quantities, and are violated if the input values are not consistent.

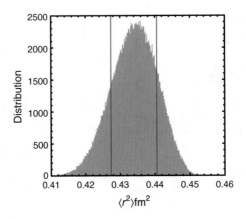

Fig. 5.6 Statistical distribution of $\langle r_\pi^2 \rangle$ for a specific experimental input. The vertical lines correspond to 68.3% CL

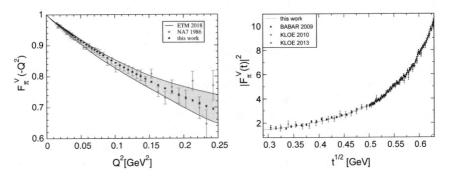

Fig. 5.7 Left: pion electromagnetic form factor in the spacelike region near the origin. Right: modulus of the form factor on the timelike axis below 0.65 GeV

0.63 GeV is compared with the experimental values of *BABAR* [42] and KLOE [44–46] experiments. The great precision of the results obtained with functional optimization methods and Monte Carlo simulations is seen in both cases.

A similar analysis of the derivative at $t = 0$ led to $\langle r_\pi^2 \rangle = (0.432 \pm 0.004)\,\text{fm}^2$, from which a very precise value of the charge radius was obtained in [56]:

$$r_\pi \equiv \sqrt{\langle r_\pi^2 \rangle} = (0.657 \pm 0.003)\,\text{fm}. \qquad (5.46)$$

Details on the specific calculations and other applications of the formalism (in particular, a precise determination of the low-energy two-pion contribution to the muon $g - 2$) can be found in [55–57].

5.5 $\pi \omega$ Form Factor

We discussed the $\pi \omega$ electromagnetic form factor in Sect. 1.3.2 as an example of violation of Schwarz reflection property. The form factor was defined in (1.69) and its discontinuity across the cut in the elastic region $4m_\pi^2 \le t < (m_\omega + m_\pi)^2$ was written in (1.70). The recent interest in this form factor is due to the discrepancies between the theoretical calculations in the standard dispersion theory [63] and the experimental data from the decay $\omega \to \pi^0 \gamma^*$ reported in [64–66].

A study of the $\pi \omega$ form factor in the Meiman-Okubo framework was performed in [67]. The purpose was to avoid the model-dependent assumptions about the discontinuity above the $\omega \pi$ threshold, adopted in the standard dispersive treatments, using instead a model-independent condition on the modulus derived from analyticity and unitarity. It turns out that for the relevant current-current correlator the lowest state allowed in the unitarity relation is the two-pion state, the opening of $\pi \omega$ channel occuring only around $1 \, \text{GeV}^2$. Therefore, unitarity leads to a relation of the form (5.8), involving the pion vector form factor F_π^V and the $\pi \omega$ form factor $f_{\pi \omega}$. We recall that in the case of the $BD^{(*)}$ form factors, heavy-quark symmetry was used in order to connect near the zero recoil the various form factors which appear in a Meiman-Okubo inequality of the form (5.8). Such symmetry does not exist in the present case. However, we can use the fact that the integral containing the pion form factor can be evaluated with a relatively good precision. Then, subtracting from both sides of (5.8) the contribution of the pion form factor, we obtain an inequality of the form

$$\frac{1}{\pi} \int_{t_{in}}^{\infty} w(t) \, |f_{\omega \pi}(t)|^2 \, dt \le I, \qquad t_{in} = (m_\omega + m_\pi)^2, \qquad (5.47)$$

where $w(t)$ is the typical Meiman-Okubo weight discussed in Sect. 5.1, and

$$I = 1 - \frac{1}{\pi} \int_{t_+}^{\infty} w(t) \, |F_\pi^V(t)|^2 \, dt. \qquad (5.48)$$

Alternatively, as shown in [68], an inequality of the type (5.47) can be derived in a straightforward way from the values of $|f_{\omega \pi}(t)|$ measured for $t > t_{in}$ from $e^+ e^- \to \omega \pi$ process [69]. In this case, the weight $w(t)$ is not a priori fixed and can be chosen such as to suppress the contribution from regions where $|f_{\omega \pi}(t)|$ is less known in order to obtain a good estimate of the upper bound I.

Recalling that $f_{\omega \pi}(t)$ has a cut in the region $t_+ \le t \le t_{in}$, with a known discontinuity given by (1.70), we can write

$$f_{\omega \pi}(t) = \frac{1}{2 \pi i} \int_{t_+}^{t_{in}} \frac{\text{disc}[f_{\omega \pi}(t')] \, dt'}{t' - t} + g(t), \qquad (5.49)$$

where $g(t)$ is analytic in the t plane cut for $t > t_{in}$. By inserting (5.49) in (5.47) and performing the conformal mapping (4.104) of the t plane cut for $t > t_{in}$ onto the unit

Fig. 5.8 Orange band: calculated allowed range of $|f_{\omega\pi}(t)/f_{\omega\pi}(0)|^2$ in the region $t < (m_\omega + m_\pi)^2$. The data are from Lepton-G [64], NA60 (2009) [65], NA60 (2011) [66] and CMD-2 (2005) [69]

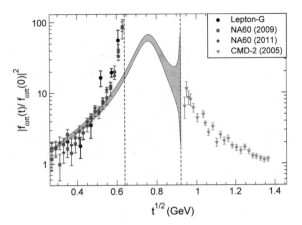

disk, we obtain a relation of the form (4.132) given in Sect. 4.5.2, i.e. a functional distance problem in L^2 norm. Then, one obtains from inequality (4.133) the allowed domain of the values $g(z_n)$ at a set of interior points. Using (5.49), these constraints are expressed in terms of the form factor, leading in particular to upper and lower bounds on the modulus $|f_{\pi\omega}(t + i\varepsilon)|$ for $t \in (t_+, t_{in})$.

However, a word of caution is in order: in Chap. 4 and in particular in Sect. 4.5.2 which we invoked here, we considered only functions of real type, which satisfy Schwarz reflection. This property is not valid for the function $f_{\omega\pi}(t)$: in particular, the Taylor coefficients g_n and the Fourier coefficients h_l of the function h are no longer real (compare the function h given in (4.113) with the dispersion integral written in (5.49)). As discussed in [67, 68], in this case the minimization should be done separately upon g_n and the complex conjugates g_n^*. The generalization is straightforward and can be found in Refs. [67, 68].

For illustration we present in Fig. 5.8 the band delimited by the upper and lower bounds on the ratio $|f_{\omega\pi}(t)/f_{\omega\pi}(0)|^2$ obtained in [68], compared with the experimental data. The results confirm the existence of a conflict between experimental data and theoretical calculations of the $\omega\pi$ form factor in the region around 0.6 GeV, and bring further arguments in support of renewed experimental efforts to measure more precisely the $\omega \to \pi^0\gamma^*$ decay.

References

1. N.N. Meiman, Zh. Eksp. Teor. Fiz. **44**, 1228 (1963) [Sov. Phys. JETP **17**, 830 (1963)]
2. S. Okubo, Phys. Rev. D **3**, 2807 (1971); Phys. Rev. D **4**, 725 (1971)
3. M. Micu, Nucl. Phys. B **44**, 531 (1972); Phys. Rev. D **7**, 2136 (1973)
4. G. Auberson, G. Mahoux, F.R.A. Simão, Nucl. Phys. B **98**, 204 (1975)
5. C. Bourrely, B. Machet, E. de Rafael, Nucl. Phys. B **189**, 157 (1981)
6. E. de Rafael, J. Taron, Phys. Lett. B **282**, 215 (1992)
7. E. de Rafael, J. Taron, Phys. Rev. D **50**, 373 (1994)

8. I. Caprini, Z. Phys. C **61**, 651 (1994)
9. I. Caprini, Phys. Lett. B **339**, 187 (1994)
10. C.G. Boyd, B. Grinstein, R.F. Lebed, Phys. Rev. Lett. **74**, 4603 (1995)
11. C.G. Boyd, B. Grinstein, R.F. Lebed, Phys. Lett. B **353**, 306 (1995)
12. C.G. Boyd, B. Grinstein, R.F. Lebed, Nucl. Phys. B **461**, 493 (1996)
13. L. Lellouch, Nucl. Phys. B **479**, 353 (1996)
14. C.G. Boyd, M.J. Savage, Phys. Rev. D **56**, 303 (1997)
15. C.G. Boyd, B. Grinstein, R.F. Lebed, Phys. Rev. D **56**, 6895 (1997)
16. I. Caprini, L. Lellouch, M. Neubert, Nucl. Phys. B **530**, 153 (1998)
17. C. Bourrely, I. Caprini, Nucl. Phys. B **722**, 149 (2005)
18. R.J. Hill, Phys. Rev. D **74**, 096006 (2006)
19. T. Becher, R.J. Hill, Phys. Lett. B **633**, 61 (2006)
20. M.C. Arnesen, J. Kundu, I.W. Stewart, Phys. Rev. Lett. **95**, 071802 (2005)
21. C. Bourrely, I. Caprini, L. Lellouch, Phys. Rev. D **79**, 013008 (2009); Erratum: Phys. Rev. D **82**, 099902 (2010)
22. G. Abbas, B. Ananthanarayan, I. Caprini, I.S. Imsong, S. Ramanan, Eur. Phys. J. A **45**, 389 (2010)
23. G. Abbas, B. Ananthanarayan, I. Caprini, I.S. Imsong, Phys. Rev. D **82**, 094018 (2010)
24. I. Caprini, E.M. Babalic, Rom. J. Phys. **55**, 920 (2010)
25. B. Ananthanarayan, I. Caprini, I.S. Imsong, Eur. Phys. J. A **47**, 147 (2011)
26. B. Grinstein, R.F. Lebed, Phys. Rev. D **92**, 116001 (2015)
27. I. Caprini, B. Grinstein, R.F. Lebed, Phys. Rev. D **96**, 036015 (2017)
28. G. Duplancic et al., JHEP **04**, 014 (2008)
29. M. Okamoto et al., Nucl. Phys. Proc. Suppl. **140**, 461 (2005)
30. E. Dalgic et al., Phys. Rev. D **73**, 074502 (2006)
31. D. Bigi, P. Gambino, Phys. Rev. D **94**, 094008 (2016)
32. D. Bigi, P. Gambino, S. Schacht, Phys. Lett. B **769**, 441 (2017)
33. D. Bigi, P. Gambino, S. Schacht, JHEP **11**, 061 (2017)
34. B. Grinstein, A. Kobach, Phys. Lett. B **771**, 359 (2017)
35. F.U. Bernlochner, Z. Ligeti, M. Papucci, D.J. Robinson, Phys. Rev. D **96**, 091503 (2017)
36. F.U. Bernlochner, Z. Ligeti, M. Papucci, D.J. Robinson, Phys. Rev. D **95**, 115008 (2017); Erratum: Phys. Rev. D **97**, 059902 (2018)
37. Y. Amhis et al., HFLAV collaboration. Eur. Phys. J. C **77**, 895 (2017)
38. L. Lellouch, E. de Rafael, J. Taron, Phys. Lett. B **414**, 195 (1997)
39. I. Caprini, Phys. Rev. D **98**, 056008 (2018)
40. R.R. Akhmetshin et al., CMD-2 collaboration. Phys. Lett. B **648**, 28 (2007)
41. M.N. Achasov et al., J. Exp. Theor. Phys. **103**, 380 (2006) [Zh. Eksp. Teor. Fiz. **130**, 437 (2006)]
42. B. Aubert et al., BaBar collaboration. Phys. Rev. Lett. **103**, 231801 (2009)
43. J.P. Lees et al., BaBar collaboration. Phys. Rev. D **86**, 032013 (2012)
44. F. Ambrosino et al., KLOE collaboration. Phys. Lett. B **670**, 285 (2009)
45. F. Ambrosino et al., KLOE collaboration. Phys. Lett. B **700**, 102 (2011)
46. D. Babusci et al., KLOE collaboration. Phys. Lett. B **720**, 336 (2013)
47. M. Ablikim et al., BESIII collaboration. Phys. Lett. B **753**, 629 (2016)
48. H. Leutwyler, in *Continuous Advances in QCD 2002*, ed. by K.A. Olive, M.A. Shifman, M.B. Voloshin (2002), pp. 23–40, arXiv:hep-ph/0212324
49. I. Caprini, Eur. Phys. J. C **13**, 471 (2000)
50. G.R. Farrar, D.R. Jackson, Phys. Rev. Lett. **43**, 246 (1979)
51. G.P. Lepage, S.J. Brodsky, Phys. Lett. B **87**, 359 (1979)
52. B. Ananthanarayan, I. Caprini, I. Sentitemsu Imsong, Phys. Rev. D **83**, 096002 (2011)
53. B. Ananthanarayan, I. Caprini, I. Sentitemsu Imsong, Phys. Rev. D **85**, 096006 (2012)
54. B. Ananthanarayan, I. Caprini, D. Das, I. Sentitemsu Imsong, Phys. Rev. D **89**, 036007 (2014)
55. B. Ananthanarayan, I. Caprini, D. Das, I. Sentitemsu Imsong, Phys. Rev. D **93**, 116007 (2016)
56. B. Ananthanarayan, I. Caprini, D. Das, Phys. Rev. Lett. **119**, 132002 (2017)

57. B. Ananthanarayan, I. Caprini, D. Das, Phys. Rev. D **98**, 114015 (2018)
58. T. Horn et al., Jefferson lab F_π collaboration. Phys. Rev. Lett. **97**, 192001 (2006)
59. G.M. Huber et al., Jefferson lab F_π collaboration. Phys. Rev. C **78**, 045203 (2008)
60. M. Schmelling, Phys. Scr. **51**, 676 (1995)
61. S.R. Amendolia et al., NA7 collaboration. Nucl. Phys. B **277**, 168 (1986)
62. C. Alexandrou et al., ETM collaboration. Phys. Rev. D **97**, 014508 (2018)
63. S.P. Schneider, B. Kubis, F. Niecknig, Phys. Rev. D **86**, 054013 (2012)
64. R.I. Dzhelyadin et al., Phys. Lett. B **102**, 296 (1981); [JETP Lett. **33**, 228 (1981)]
65. R. Arnaldi et al., NA60 collaboration. Phys. Lett. B **677**, 260 (2009)
66. G. Usai [NA60 collaboration], Nucl. Phys. A **855**, 189 (2011)
67. B. Ananthanarayan, I. Caprini, B. Kubis, Eur. Phys. J. C **74**, 3209 (2014)
68. I. Caprini, Phys. Rev. D **92**, 014014 (2015)
69. R.R. Akhmetshin et al., CMD-2 collaboration. Phys. Lett. B **562**, 173 (2003)

Chapter 6
Functional Analysis and Optimization Methods in Perturbative QCD

In this chapter we first show that ideas from hyperasymptotics can be used to infer the existence of terms additional to the usual operator product expansions (OPE) of QCD correlators. A connection between these quark-hadron duality-violating terms and the singularities of the Borel-Laplace transform in the Borel plane is emphasized. We then discuss a method based on functional analysis for testing the presence of duality violation from experimental data. Finally, an alternative way to go beyond finite-order perturbation theory, by a modified perturbative expansion based on the conformal mapping of the Borel plane, is briefly discussed.

6.1 Divergent Expansions and Hyperasymptotics in QCD

As we discussed in Sect. 2.3, the application of perturbative QCD to hadronic processes is based on the concept of quark-hadron duality. In its modern formulation, quark-hadron duality assumes that the description of the Green functions in the frame of Wilson's operator product expansion (OPE), valid away from the Minkowski axis, can be analytically continued to match with the description in terms of hadrons, which live on the Minkowski axis.

We consider as in Sect. 2.3 the QCD Adler function $D(s)$ defined by (2.7), where the polarization function $\Pi(s)$ is analytic in the s plane cut for $s \geq s_+$ and satisfies the dispersion relation (2.9).

From (2.14) and (2.18), the OPE representation of the function D in our normalization reads

$$D_{\text{OPE}}(s) \sim 1 + \sum_{n \geq 1} c_{n,1}\, a_s^n + \sum_{k \geq 1} \frac{d_k(s)\langle O_k \rangle}{(-s)^k}, \quad a_s \equiv \frac{\alpha_s(-s)}{\pi}, \qquad (6.1)$$

where $\alpha_s(-s)$ is the strong coupling given to one loop by (2.15), $\langle O_k \rangle$ denote nonperturbative condensates of increasing dimensionality and the coefficients $d_k(s)$ depend logarithmically on s.

© The Author(s), under exclusive licence to Springer Nature Switzerland AG 2019
I. Caprini, *Functional Analysis and Optimization Methods in Hadron Physics*,
SpringerBriefs in Physics, https://doi.org/10.1007/978-3-030-18948-8_6

Both expansions in (6.1) are expected to be divergent series, with zero convergence radii (we indicated this by the sign $\sim$ used conventionally for asymptotic series). The zero convergence radius follows from the fact that the expanded function is singular at the expansion point, $|s| \to \infty$, as a consequence of the dispersion relation (2.9) satisfied by $\Pi(s)$ and the accumulation of the hadronic branch points in the spectral function at infinity, predicted by unitarity. In particular, using renormalization group invariance, 't Hooft [1] and Khuri [2] expressed the singularities of $D(s)$ at large s as curves of singularities accumulating at the origin of the complex plane of the coupling. The singularity at $a_s = 0$ indicates that the perturbative expansion in (6.1) should be divergent. This property is suggested alternatively from explicit calculations of Feynman diagrams, which give a factorial growth of the expansion coefficients, $c_{n,1} \sim n!$ (for a review and references see [3]). The properties of the second series in (6.1) are less known, however it is believed that it is too a divergent series. Moreover, there are strong arguments that the expansion (6.1) is not complete. As discussed in a series of papers [4–11], additional terms, beyond OPE, are present in the expansions of the QCD Green functions. Traditionally, these terms are known as "quark-hadron duality-violating" (DV) terms.

The existence of terms that should be added to (6.1) is suggested by mathematical theories of hyperasymptotics and transseries [12]. Given an asymptotic perturbation series of a function $F(g)$ in powers of a small coupling g:

$$F(g) \sim \sum_{n \geq 0} F_n \, g^n, \qquad g \to 0_+, \tag{6.2}$$

one can add in the r.h.s. a term exponentially small in the coupling and write

$$F(g) \sim \sum_{n \geq 0} F_n \, g^n + O(\exp[-c/g]), \qquad g \to 0_+, \tag{6.3}$$

where $c > 0$ is arbitrary, because all the perturbative coefficients of the added term are zero. More generally, one can add a whole series in the parameter $\exp[-c/g]$, a so-called "transseries". In this respect, the series of power corrections in (6.1) can be viewed as a transseries of the purely perturbative expansion in powers of a_s, since, using (2.15), we have $\exp[-c/\alpha_s(-s)] \sim (\Lambda^2/(-s))^c$. Going further, the transseries associated to the series of power corrections contains terms of the form $\exp[-\xi s]$, for a certain constant ξ. Thus, one expects the duality-violating terms to be exponentially decreasing at large $|s|$ [11]. However, it is impossible to construct an analytic function with an exponential decrease in all directions of the complex plane: a function that decreases exponentially on the Euclidean axis $s < 0$ gives by analytic continuation a function that either increases exponentially or oscillates on the Minkowski axis $s > 0$. A more detailed analysis is therefore necessary in order to understand the origin and the properties of the DV terms.

From simple models [5, 9], one expects the additional terms to OPE to be non-zero only in the right half $\text{Re } s > 0$ of the s-plane, in particular on the timelike axis $s > 0$. Strong arguments in favour of this structure of the DV terms have been brought

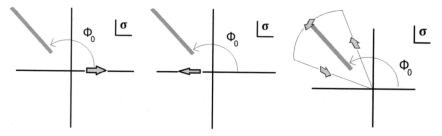

Fig. 6.1 Rotation of the integration axis from the positive semiaxis to the negative semiaxis in the σ plane. Blue line: branch cut expected when the spectral function consists from an infinite set of resonances in the limit of large, but finite N_c. Last panel: contour around the additional cut

recently in [13], from a study based on the analytic continuation of the generalized Borel-Laplace transform $\mathscr{B}(\sigma)$, defined as

$$\mathscr{B}(\sigma) = \int_0^\infty dt \, \mathrm{Im} \, \Pi(t) \, e^{-\sigma t}, \tag{6.4}$$

in terms of the spectral function of the polarization function. The usefulness of this function is seen by noting that from the definition (2.7) of the Adler function and the dispersion relation (2.9) one obtains the representation

$$D(s) = -s \int_0^\infty d\sigma \, e^{\sigma s} \, \sigma \mathscr{B}(\sigma), \tag{6.5}$$

valid in the left half $\mathrm{Re}\, s < 0$ of the s-plane. This representation is suitable for investigating the expansion of $D(s)$ at large $|s|$. In particular, the expansion of $\mathscr{B}(\sigma)$ near $\sigma = 0$ generates power-suppressed terms at large s.

The main remark made in [13] is that the representation (6.5), valid originally in the half plane $\mathrm{Re}\, s < 0$ of the s complex plane, can be analytically continued to regions where $\mathrm{Re}\, s > 0$ by rotating the integration axis in (6.5) from the positive semiaxis to the negative semiaxis of the σ-plane. If some singularities of $\mathscr{B}(\sigma)$ are met in this rotation, they will give rise, by Cauchy theorem, to new terms in the expansion of the Adler function. Thus, a rigorous connection was established in [13] between the singularities of the Borel transform $\mathscr{B}(\sigma)$ in the σ-plane and the asymptotic expansion of the Adler function. As discussed in detail in [13] on a class of models for the spectral function $\mathrm{Im}\Pi(t)$, the behavior of $\mathscr{B}(\sigma)$ near $\sigma = 0$ is correlated with the OPE, while the singularities away from the origin are responsible for the appearance of DV terms.

For illustration we show in Fig. 6.1 the σ plane for the most complex case investigated in [13], when the spectral function $\mathrm{Im}\Pi(t)$ contains an infinite set of resonances in the limit of large, but finite number N_c of colours, and the Borel transform is expected to have a branch cut (indicated by the blue line) in the left half of the σ plane. The integral around this cut shown in the last panel of Fig. 6.1, generates a new,

DV term, which is nonzero only for Re $s > 0$, and can be written, up to logarithmic corrections, as

$$\Pi_{\mathrm{DV}}(s) \sim C e^{-\xi s + i\eta s}, \quad \mathrm{Re}\, s > 0, \quad \mathrm{Im}\, s \geq 0, \tag{6.6}$$

where C, $\xi > 0$ and η are real numbers related to the parameters which describe the resonance spectrum. The expression for Im $s \leq 0$ is obtained using the Schwarz reflection $\Pi_{\mathrm{DV}}(s^*) = \Pi_{\mathrm{DV}}^*(s)$. As discussed in [13], in the left half of s plane, in particular on the Euclidean axis, one expects DV terms which decrease exponentially faster than those in the right half plane. Therefore, the DV terms can be neglected on the Euclidean axis. On the other hand, the DV terms (6.6) are expected to produce visible effects on the Minkowski axis. Methods to detect DV from experimental data are discussed in the next section.

6.2 Tests of Quark-Hadron Duality Violation

In Sect. 2.3, we briefly reviewed the standard method of testing OPE by means of the finite-energy sum rules (FESR) and brought arguments in favour of an approach based on functional analysis techniques. In Sect. 2.4.2, we formulated the extremal problem (2.26), which gives a lower bound on the functional distance between the admissible class containing the true function and the approximate expansion given by perturbative QCD. In what follows we shall show how this extremal problem can be used for testing the presence of DV terms in the asymptotic expansion of a QCD correlator from error-affected experimental data on its spectral function at low energies.

As in Refs. [14, 15], we consider the exact model proposed in Refs. [6, 9], where the correlator $\Pi(s)$ has the closed expression

$$\Pi_{\mathrm{model}}(s) = -\frac{1}{\zeta} \frac{2F^2}{\Lambda^2} \psi\left(\frac{v + m_0^2}{\Lambda^2}\right), \tag{6.7}$$

with

$$v = \Lambda^2 \left(\frac{-s - i\varepsilon}{\Lambda^2}\right)^{\zeta}. \tag{6.8}$$

We recall that $\psi(v) = \Gamma'(v)/\Gamma(v)$ is the Euler digamma function and the parameters ζ, F, Λ and m_0 are fixed such as the spectral function

$$\sigma(s) = \mathrm{Im}\, \Pi_{\mathrm{model}}(s + i\varepsilon), \quad 0 \leq s \leq s_0, \tag{6.9}$$

provides a realistic description of the vector spectral function of the QCD correlator at low energies, measured experimentally from hadronic τ decays (for instance, by ALEPH experiment [16]).

At large s, the correlator $\Pi_{\mathrm{model}}(s)$ can be approximated by a truncated OPE-type expansion

$$\Pi_{\mathrm{model}}(s) \sim \Pi_{\mathrm{OPE}}(s) = -\frac{2F^2}{\Lambda^2} C_0 \log\left(\frac{-s}{\Lambda^2}\right) + \sum_{k=1}^{N_{\mathrm{OPE}}} \frac{C_{2k}}{v^k}, \qquad (6.10)$$

where the logarithmic term corresponds to the "purely perturbative" part and the other terms are power corrections, with coefficients C_{2k} expressed in terms of Bernoulli polynomials $B_k(x)$ [9]. In addition, the asymptotic expansion of the model contains a DV contribution, which vanishes for $\mathrm{Re}(s) \leq 0$ and is given for $\mathrm{Re}(s) > 0$ and $\mathrm{Im}(s) > 0$ by

$$\Pi_{\mathrm{DV}}(s) = \frac{2\pi F^2}{\Lambda^2 \zeta}\left[-i + \cot\left[\pi\left(\frac{-s}{\Lambda^2}\right)^{\zeta} + \pi\frac{m_0^2}{\Lambda^2}\right]\right]. \qquad (6.11)$$

For $\mathrm{Im}(s) < 0$, the DV term is obtained by Schwarz reflection $\Pi_{\mathrm{DV}}(s^*) = \Pi_{\mathrm{DV}}^*(s)$.

The question of interest is to what extent the low-energy data on $\sigma(s)$ feel the presence of the DV terms in the asymptotic expansion of the model. The test performed in [14, 15] is based on the auxiliary function

$$\Pi_\mu(s) = \Pi_{\mathrm{OPE}}(s) + \mu\Pi_{\mathrm{DV}}(s), \qquad (6.12)$$

depending on a "strength" parameter μ which rescales the magnitude of the DV term. The problem is whether and with how much precision one can detect the true value $\mu = 1$ of this parameter from error-affected experimental data on the spectral function (6.9) available for $s \leq s_0$.

By a simple change of variable $z = s/s_0$, the extremal problem (2.26) formulated in Sect. 2.4.2 can be cast in the form of a minimum-distance problem in L^∞ norm, which we write as

$$\delta_\infty(\mu) = \min_{g \in H^\infty} \|g - h\|_\infty, \qquad (6.13)$$

where H^∞ denotes the class of functions analytic and bounded in $|z| < 1$ and the complex function h is

$$h(\zeta) = -\frac{1}{\pi}\int_0^1 \frac{\sigma(s_0 x)\,dx}{x - \zeta} + \Pi_{\mathrm{OPE}}(s_0\zeta) + \mu\Pi_{\mathrm{DV}}(s_0\zeta). \qquad (6.14)$$

The problem (6.13) coincides actually with the extremal problem (4.134) solved in Sect. 4.6. We recall that the exact solution of this problem is given in Eq. (4.143), while an approximate solution based on a suitable class of weighted L^2 norms is presented in Sect. 4.7.

Having in view the meaning of the quantity δ_∞ as the minimum distance from the admissible class containing the physical function $\Pi(s)$ to the approximant $\Pi_\mu(s)$, it

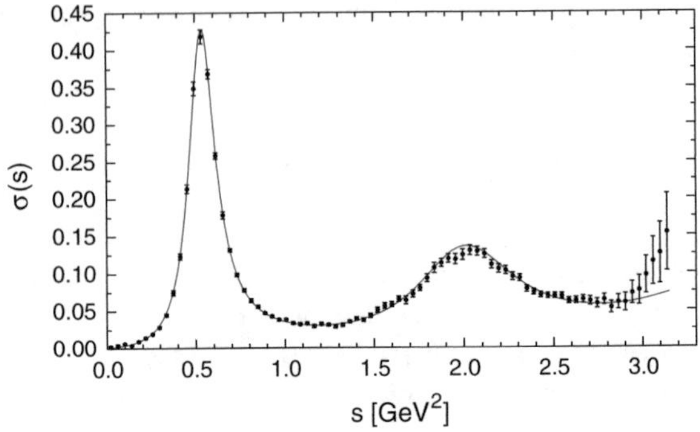

Fig. 6.2 Example of a data set on the spectral function (6.9), obtained from the central values of the model given in Eq. (6.7) with covariances from a numerical interpolation of ALEPH's covariance matrix for the vector channel [16]. The solid line gives the central value of the model

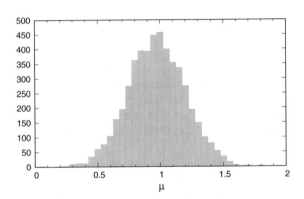

Fig. 6.3 Typical μ distribution obtained from Monte Carlo simulations

is reasonable to take as optimal μ the value which minimizes δ_∞ for each input on the spectral function. In order to further simulate the experimental situation met in the application of perturbative QCD to hadronic τ decays (see for instance [17–20] and references therein), the upper limit s_0 was taken equal to or lower than m_τ^2, and fake data on the spectral function have been generated in a number of bins, using a multivariate Gaussian distribution with covariances inferred from the covariance matrix published by ALEPH experiment [16]. An example of a data set generated in this way in Ref. [13] is displayed in Fig. 6.2. In this figure, the strong correlations between data points are clearly visible, mainly towards the end-point of the spectrum, where the uncertainties are also larger.

By simulations with a large number of different data sets, a statistical distributions of the optimal parameter μ was obtained. A typical distribution obtained in [13], shown in Fig. 6.3, is seen to be to a very good approximation Gaussian, allowing the extraction of a mean value and a standard deviation.

The detailed analysis performed in [15] showed that for this particular model the true value $\mu = 1$ of the strength parameter was obtained with high accuracy from error-affected data on the spectral function. The next step is of course the application of the method to real data and to real QCD, using the insight on the general structure of the DV terms gained from the analysis presented in the previous section.

6.3 Conformal Mappings of the Borel Plane

In this section we shall briefly discuss an alternative way to go beyond finite-order perturbation theory for improving the predictions of perturbative QCD. It is based on the use of conformal mappings for enlarging the domain of convergence and for improving the convergence rate of power series.

We introduced this method in Sect. 4.9, where we defined in particular the optimal conformal mapping which leads to the best asymptotic convergence rate. As remarked at the end of Sect. 4.10, in QCD this method cannot be applied to the perturbative expansions in powers of the strong coupling α_s, because, as argued in [1, 2], the expanded function is singular at the expansion point $\alpha_s = 0$. However, the method of conformal mappings can be applied to the Borel plane, for improving the expansions of the Borel transforms of the QCD correlators.

Starting from the perturbative expansion the Adler function (the first sum in (6.1)), we define the Borel transform $B(u)$ by

$$B(u) = \sum_{n=0}^{\infty} b_n u^n, \qquad b_n = \frac{c_{n+1,1}}{\beta_0^n \, n!}, \tag{6.15}$$

where β_0 is the first coefficient of the renormalization-group β function (we recall that $\beta_0 = 9/4$ for $n_f = 3$, as mentioned below Eq. (2.15)). Then the purely perturbative expansion in the OPE representation (6.1) can be written formally in terms of $B(u)$ by means of the Laplace-Borel representation

$$D_{\mathrm{pert}}(s) = 1 + \frac{1}{\beta_0 a_s} \int_0^{\infty} du \, B(u) \, \exp\left(-\frac{u}{\beta_0 a_s}\right), \tag{6.16}$$

which reproduces term by term the original expansion (6.1).

As it is known, the large-order increase of the coefficients $c_{n,1}$ of the perturbation series is encoded in the singularities of the Borel transform $B(u)$ in the complex u plane [3, 21, 22]. The Borel transform of the Adler function has singularities situated on the semiaxis $u \geq 2$, known as infrared (IR) renormalons, and on the semiaxis $u \leq -1$, known as ultraviolet (UV) renormalons (see Fig. 6.4, left).

Due to the singularities of $B(u)$ for $u \geq 2$, the Laplace-Borel integral (6.16) is ambiguous and requires a regularization.[1] We shall adopt the principal value (PV) prescription [3, 21], defined as the semisum of integrals of the form (6.16) along lines C_+ (C_-) parallel to the positive real axis, slightly displaced above (below) it. As discussed in [23], the PV prescription is suitable from the point of view of momentum-plane analyticity, in particular, it preserves Schwarz reflection property in the complex s plane.

---

[1]In mathematical language, we say that the Adler function is not Borel summable. The Borel nonsummability results also from the structure of the singularities of D near the origin of the a_s complex plane, obtained in [1, 2].

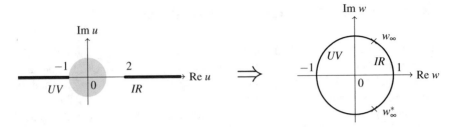

Fig. 6.4 Left: Borel plane of the Adler function. The gray circle is the convergence domain of the series (6.15). Right: the w plane obtained by the conformal mapping (6.17). The IR and UV renormalons are mapped on the boundary of the unit disk. The points w_∞ (w_∞^*) are the images of $u \to \infty$ in the upper (lower) half planes

As shown in Fig. 6.4 left, the convergence of the series (6.15) is limited to the disk $|u| < 1$ imposed by the first renormalon at $u = -1$. A conformal mapping which enlarges the convergence domain by taking into account this singularity was proposed in [21]. However, this mapping is not optimal according to the definition given in Sect. 4.9. The optimal conformal mapping of the Borel plane in the case of the Adler function, found in Ref. [24], is a particular case of the mapping (4.3) given in Sect. 4.1. It has the form

$$w \equiv \tilde{w}(u) = \frac{\sqrt{1+u} - \sqrt{1-u/2}}{\sqrt{1+u} + \sqrt{1-u/2}}, \tag{6.17}$$

and maps the complex u-plane cut along the real axis for $u \geq 2$ and $u \leq -1$ onto the interior of the circle $|w| < 1$ in the complex w-plane, such that $\tilde{w}(0) = 0$ (see Fig. 6.4, right). Therefore, the expansion of $B(u)$ in powers of the variable w:

$$B(u) = \sum_{n=0}^{\infty} c_n w^n, \tag{6.18}$$

converges in the whole disk $|w| < 1$ and hence in the whole u-plane up to the cuts.

By formally inserting the series (6.18) into (6.16), one is led to consider an expansion of the form [24–26]:

$$D_{\text{pert}}(s) = 1 + \sum_{n=0}^{\infty} c_n W_n(a_s), \tag{6.19}$$

where the functions $W_n(a_s)$ are defined as

$$W_n(a_s) = \frac{1}{\beta_0 a_s} \text{PV} \int_0^{\infty} e^{-u/(\beta_0 a_s)} (\tilde{w}(u))^n \, du. \tag{6.20}$$

At each finite truncation order N, the expansion (6.19) is a trivial reordering of the original series (6.1). For $N \rightarrow \infty$, however, the new expansion (6.19) represents a nontrivial step out of perturbation theory, replacing the perturbative powers a_s^n by the functions $W_n(a_s)$.

It is possible to exploit in addition the fact that the nature of the leading singularities in the Borel plane is known: near the first branch points, $u = -1$ and $u = 2$, $B(u)$ behaves like

$$B(u) \sim \frac{r_1}{(1+u)^{\gamma_1}} \quad \text{and} \quad B(u) \sim \frac{r_2}{(1-u/2)^{\gamma_2}}, \qquad (6.21)$$

respectively, where the residues r_1 and r_2 are not known, but the exponents γ_1 and γ_2 can be calculated and have known values [3, 21, 27].

As discussed in [24–26], while the optimal conformal mapping is unique, the implementation of the singular behavior (6.21) can be done in various ways, which influence the low-order properties of the series, but lead to the same predictions at large orders. One possibility is the expansion

$$D_{\mathrm{pert}}(s) = 1 + \sum_{n=0}^{\infty} \hat{c}_n \widehat{W}_n(a_s), \qquad (6.22)$$

where the functions $\widehat{W}_n(a_s)$ are defined as

$$\widehat{W}_n(a_s) = \frac{1}{\beta_0 a_s} \mathrm{PV} \int_0^{\infty} e^{-u/(\beta_0 a_s)} \frac{(\tilde{w}(u))^n}{(1+u)^{\gamma_1}(1-u/2)^{\gamma_2}} \, du. \qquad (6.23)$$

Other expansion functions have been studied in [28–33], where the good convergence of the new expansions (6.19) and (6.22) have been demonstrated by detailed analyses for simple mathematical models [17] proposed for the Adler function.

On a more fundamental level, the new expansions (6.19) and (6.22) have interesting properties which make them good candidates for a modified QCD perturbation theory. By construction, when reexpanded in powers of a_s, these series reproduce the perturbative part of the expansion (6.1) with the coefficients $c_{n,1}$ known from Feynman diagrams. The expansion functions $W_n(a_s)$ and $\widehat{W}_n(a_s)$ are well defined and have bounded magnitudes for Re $a_s > 0$. However, they are singular at $a_s = 0$ and their perturbative expansions in powers of a_s are divergent series, resembling from this point of view the expanded function D itself. For illustration, we give below the expansions of the first functions $W_n(a_s)$ defined in (6.20):

$$W_1(a_s) \sim 0.844\, a_s - 0.947\, a_s^2 + 5.206\, a_s^3 - 27.68\, a_s^4 + 249.11\, a_s^5 + \cdots \quad (6.24)$$
$$W_2(a_s) \sim 1.423\, a_s^2 - 4.807\, a_s^3 + 40.49\, a_s^4 - 334.46\, a_s^5 + 3863.9\, a_s^6 + \cdots$$
$$W_3(a_s) \sim 3.599\, a_s^3 - 24.32\, a_s^4 + 290.63\, a_s^5 - 3366.9\, a_s^6 + 49041.2\, a_s^7 + \cdots$$

On the other hand, the new expansions (6.19) and (6.22) have a tamed behavior at high orders and, under certain conditions, they may even converge in a domain of the s-plane [25]. Further studies are necessary in order to understand the connection between the method of conformal mappings presented in this section and the approach based on hyperasymptotics and transseries, discussed in Sect. 6.1.

References

1. G. 't Hooft, in *The Whys of Subnuclear Physics*, ed. by A. Zichichi (Plenum Press, New York, 1979)
2. N.N. Khuri, Phys. Lett. **82B**, 83 (1979)
3. M. Beneke, Phys. Rept. **317**, 1 (1999)
4. M.A. Shifman, Int. J. Mod. Phys. A **11**, 3195 (1996)
5. B. Blok, M.A. Shifman, D.X. Zhang, Phys. Rev. D **57**, 2691 (1998); Erratum: Phys. Rev. D **59**, 019901 (1999)
6. M.A. Shifman, in *At the Frontier of Particle Physics*, vol. 3 (World Scientific, Singapore, 2001), pp. 1447–1494
7. M. Golterman, S. Peris, B. Phily, E. de Rafael, JHEP **01**, 024 (2002)
8. O. Catá, M. Golterman, S. Peris, JHEP **08**, 076 (2005)
9. O. Catá, M. Golterman, S. Peris, Phys. Rev. D **77**, 093006 (2008)
10. O. Catá, M. Golterman, S. Peris, Phys. Rev. D **79**, 053002 (2009)
11. M. Shifman, J. Exp. Theor. Phys. **120**, 386 (2015)
12. See, for instance, M.V. Berry, C.J. Howls, Proc. R. Soc. Lond. A **430**, 653 (1990); J.P. Boyd, Acta Applic. Mathem. **56**, 1 (1999)
13. D. Boito, I. Caprini, M. Golterman, K. Maltman, S. Peris, Phys. Rev. D **97**, 054007 (2018)
14. I. Caprini, M. Golterman, S. Peris, Phys. Rev. D **90**, 033008 (2014)
15. D. Boito, I. Caprini, Phys. Rev. D **95**, 074027 (2017)
16. M. Davier, A. Höcker, B. Malaescu, C.Z. Yuan, Z. Zhang, Eur. Phys. J. C **74**, 2803 (2014)
17. M. Beneke, M. Jamin, JHEP **09**, 044 (2008)
18. D. Boito, O. Catá, M. Golterman, M. Jamin, K. Maltman, J. Osborne, S. Peris, Phys. Rev. D **84**, 113006 (2011)
19. A. Pich, A. Rodríguez-Sánchez, Phys. Rev. D **94**, 034027 (2016)
20. D. Boito, M. Golterman, K. Maltman, S. Peris, Phys. Rev. D **95**, 034024 (2017)
21. A.H. Mueller, in *QCD; 20 Years Later*, vol. 1, ed. by P. Zerwas, H.A. Kastrup (World Scientific, Singapore, 1993)
22. A.H. Mueller, Nucl. Phys. B **250**, 327 (1985)
23. I. Caprini, M. Neubert, JHEP **03**, 007 (1999)
24. I. Caprini, J. Fischer, Phys. Rev. D **60**, 054014 (1999)
25. I. Caprini, J. Fischer, Phys. Rev. D **62**, 054007 (2000)
26. I. Caprini, J. Fischer, Eur. Phys. J. C **24**, 127 (2002)
27. M. Beneke, V.M. Braun, N. Kivel, Phys. Lett. B **404**, 315 (1997)
28. I. Caprini, J. Fischer, Eur. Phys. J. C **64**, 35 (2009)
29. I. Caprini, J. Fischer, Phys. Rev. D **84**, 054019 (2011)
30. G. Abbas, B. Ananthanarayan, I. Caprini, J. Fischer, Phys. Rev. D **87**, 014008 (2013)
31. G. Abbas, B. Ananthanarayan, I. Caprini, J. Fischer, Phys. Rev. D **88**, 034026 (2013)
32. I. Caprini, J. Fischer, G. Abbas, B. Ananthanarayan, in *Perturbation Theory; Advances in Research and Applications* (Nova Science Publishers, 2018), pp. 211–254
33. I. Caprini, Phys. Rev. D **98**, 056016 (2018)

Printed in the United States
By Bookmasters